Weigang Zhang
Signals and Systems
De Gruyter Graduate

Also of Interest

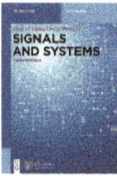

Signals and systems
G. Li, L. Chang, S. Li, 2015
ISBN 978-3-11-037811-5, e-ISBN (PDF) 978-3-11-037954-9, e-ISBN
(EPUB) 978-3-11-041684-8

Energy harvesting
O. Kanoun (Ed.), 2017
ISBN 978-3-11-044368-4, e-ISBN (PDF) 978-3-11-044505-3, e-ISBN
(EPUB) 978-3-11-043611-2, Set-ISBN 978-3-11-044506-0

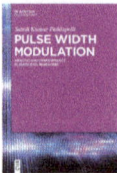

Pulse width modulation
S. Peddapelli, 2016
ISBN 978-3-11-046817-5, e-ISBN (PDF) 978-3-11-047042-0, e-ISBN
(EPUB) 978-3-11-046857-1, Set-ISBN 978-3-11-047043-7

Wind energy & system engineering
D. He, G. Yu, 2017
ISBN 978-3-11-040143-1, e-ISBN 978-3-11-040153-0, e-ISBN (EPUB)
978-3-11-040164-6, Set-ISBN 978-3-11-040154-7

Weigang Zhang

Signals and Systems

Volume 2: In discrete time

DE GRUYTER

清华大学出版社
TSINGHUA UNIVERSITY PRESS

Author
Prof. Weigang Zhang
Chang' an University
Mid South 2nd Ring Road
Shaanxi Province
710064 XI 'AN China
wgzhang@chd.edu.cn; 648383177@qq.com

ISBN 978-3-11-054118-2
e-ISBN (PDF) 978-3-11-054120-5
e-ISBN (EPUB) 978-3-11-054124-3

Library of Congress Cataloging-in-Publication Data
A CIP catalog record for this book has been applied for at the Library of Congress.

Bibliographic information published by the Deutsche Nationalbibliothek
The Deutsche Nationalbibliothek lists this publication in the Deutsche Nationalbibliografie;
detailed bibliographic data are available on the Internet at http://dnb.dnb.de.

© 2018 Walter de Gruyter GmbH, Berlin/Boston
Cover image: Creats/Creatas/thinkstock
Typesetting: le-tex publishing services GmbH, Leipzig
Printing and binding: CPI books GmbH, Leck
♾ Printed on acid-free paper
Printed in Germany

www.degruyter.com

Preface

The Signals and Systems course is an important professional fundamental course for undergraduates majoring in electronics, information and communication, and control, etc. It has a profound influence on the cultivation of students' overall abilities, such as independent learning, scientific thinking in problem solving, practical skills, etc. The course is not only compulsory for undergraduates but also necessary for postgraduate entrance examination in related majors. The course plays a critical role in undergraduate education, and it is the theoretical foundation for information theory and mastering information technology. It is even regarded as the key to opening the door to information science in the twenty-first century.

This book is Volume 2 of "Signals and Systems" written by Weigang Zhang and published by De Gruyter Press and Tsinghua University Press for the undergraduate course "Signals and Systems". The problems about discrete time signals and systems analysis in the time domain and the z-domain and the state-space analysis of systems were mentioned in this volume. At the end of the volume, several examples in communication systems were given, which relate to signals and systems analysis theories and methods.

I would like to express my sincere thanks to Associate Professor Wei-Feng Zhang, the coauthor of this book, and Associate Professors Tao Zhang and Shuai Ren, and lecturers Xiao-Xian Qian and Jian-Fang Xiong for preparing the Solved Questions section and translating the Chinese manuscript into English. I also wish to thank Dan-Yun Zheng, Jie-Xu Zhao, Xiang-Yun Li, Pei-Cheng Wang, Juan-Juan Wu, Jing Wu and Run-Qing Li for proofreading all examples, exercises and answers and helping with the translation of the manuscript. Finally, thanks are also due to authors and translators of reference books in the book.

The books are a summary of the authors' many years of teaching experience; any suggestions that would help improve the book would be greatly appreciated. Please feel free to contact us for any problems encountered during reading; we would grateful for your comments.

June 2016

Weigang Zhang

https://doi.org/10.1515/9783110541205-201

Preface of Volume I

The course of *Signals and systems* is an important professional, fundamental course for undergraduates majoring in electronics, information and communication, and control, etc. The course has a profound influence on the cultivation of students' over-all capabilities, such as independent learning, scientific thinking in problem solving, practical skills, etc. The course is not only compulsory for undergraduates, but it is also necessary for postgraduate entrance examinations in related majors. The course plays a critical role in undergraduate education and it is the theoretical foundation to information theory and information technology. It is even regarded as the key to opening the door to information science in the twenty-first century.

Main contents in the course

From the content aspect, the course of Signals and Systems is more of a mathematics course integrated into professional characteristics than a specialized course. The so called signal is actually the function in mathematics, given "voltage", "current" or other physical background. System is considered as a module that can transfer (process) a signal.

In essence, the contents of the course can be summarized as the study of the relation between before and after a signal is transformed by a given system or a function is processed by a given operation module. Here, the signal before being transformed is called the input or the excitation, and the signal after being transformed is called the output or the response. The excitation is the cause (or the independent variable in mathematical description), and the response is the consequence (or the dependent variable in mathematical description). Mathematically, it can be described as after an independent variable (excitation) is calculated by a module (system) a dependent variable (response) is the result (plotted in ▸ Figure 1). A real physical system (a transformer and the relation between voltages on its two ports) is shown in ▸ Figure 1a and the mathematical description or the equivalent model of this physical system is given

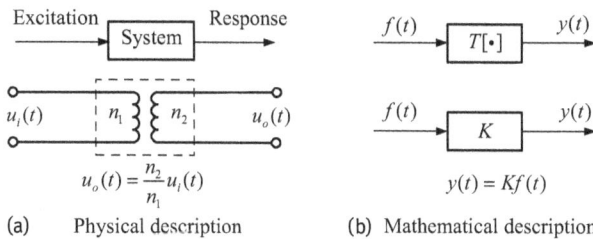

(a) Physical description

(b) Mathematical description

Fig. 1: Excitation, response and system.

https://doi.org/10.1515/9783110541205-202

by ▸ Figure 1b. From the figures it can be seen that, in fact, Signals and Systems is just a course that abstracts a physical system as a mathematical model, and then studies the performance of the system through analyzing the model, namely solving the relationship between the excitation and the response. Please note that the symbol $T[\bullet]$ in ▸ Figure 1 shows a processing method or a transformation to the element in brackets [ ]. It is obvious that the signal is an object processed by the system, and the system is the main carrier to process the signal; both complement each other.

Herein, the relationship between signal and system is described by a mathematical model (mathematical analysis formula), such as $y(t) = T[f(t)]$, therefore, solving the mathematical model or solving the equation in layman's terms will run through the course. The various solutions for the model are main knowledge points of the book.

Since the signal and the system are the two key points of this course, all research is concentrated around them.

The analysis of signals covers the following points:

(1) Modeling. Various physical signals in the real world can be mathematically abstracted into mathematical models, and the process is known as "mathematical modeling." The aim of the modeling is to change a physical signal into a function that can be analyzed theoretically on paper.

(2) Decomposition and composition of signals. One signal can be decomposed into a linear combination of other signals; or, a set of signals can be used to represent a signal with their linear combination.

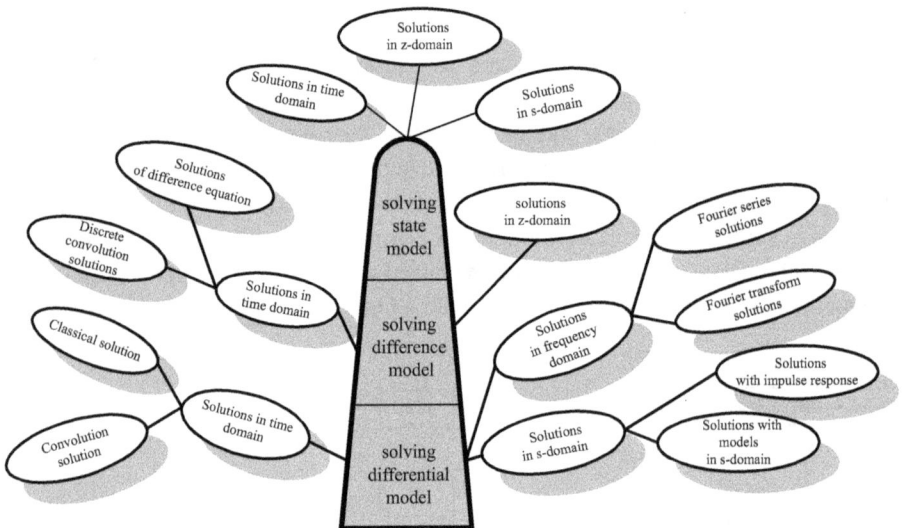

Fig. 2: Tree of content in Signals and Systems.

The analysis of systems focuses on the study of the response of a given system to an arbitrary excitation (▶ Figure 2), or, analyzes the transform characteristics of a system to signals known under the system constitution.

Briefly, this book consists of two parts, signal analysis and system analysis, and discusses how to solve differential or difference equations in the time and transform domains (real frequency domain, complex frequency domain and z domain).

Features of the course

This course has three main features:

(1) A sound theoretical basis. Various mathematical methods to solve the differential or difference equations in the time and transform domains are introduced.
(2) A strong specialty. Building up mathematical models of various systems from the real world must rely on the fundamental laws and theorems of related fields.
(3) Wide applications. The research results can be generalized to real applications in nature and society, even to nonlinear system analysis.

The aims of learning

After thinking carefully, we find that the real world is constituted by various systems. For example, the human body includes the nervous, blood and digestion systems, etc.; and then there is transport, lighting, water supply, finance, communication, control systems and so on in daily life. The functions of systems can be summarized as processing or transforming an input. So, the relationships between the inputs and the outputs of these systems are exactly the main topic studied in this book.

To facilitate research, real physical systems can be abstracted into mathematical models. Further, these models can be classified into two types of linear and nonlinear systems according to their characteristics. Thus, the aims of learning here are to master methods to analyze the relationship between the excitation and the response of a linear system and to apply these analytical results to the analysis of nonlinear systems and then to solve various practical problems in real systems.

Learning the course can help us to establish a correct, scientific and reasonable approach to analyzing and solving problems, and to improve the treatment ability of various problems and difficulties encountered in study, work and life, while at the same time, to master how to solve practical problems using basic knowledge, especially mathematical knowledge.

Research route of this book

The contents of this book can be divided into two layers: the lower layer which is signal analysis and the upper layer which is system analysis. The lower layer is the basis of the upper layer, while the upper one is the achievement of the lower one. Based on ▸ Figure 2, the research route of this book is shown in ▸ Figure 3.

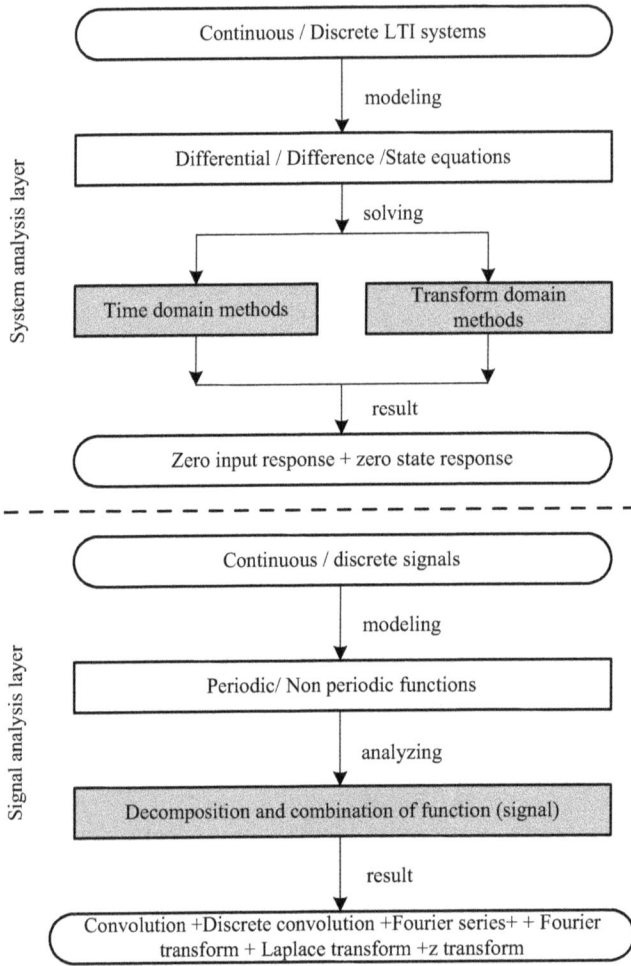

Fig. 3: The roadmap of the book.

Relationship between the course and other basic courses

There is no doubt that mathematics is an important basic course for signals and systems. The mathematics knowledge herein includes expansion and summation of series, solutions of differential/difference and algebraic equations, partial fraction expansion, basic calculus operations and linear algebra. In addition, this book, which takes electric systems as the objects to be studied, also involves professional knowledge of circuit analysis, analog circuits and digital circuits. Among of these fields, circuit analysis relates closely to the work herein; it is the precursor of the course, and the course is an expansion of content and the improvement in methodologies of the former. The similarities and differences between them are listed below:

(1) Both of the objects to be studied are circuits or networks consisting of electronic components.

(2) Both of the main aims are to find circuit variables such as voltage and current.

(3) The main analysis method in circuit analysis is to obtain the responses (node voltages and branch currents) of a circuit to excitations by constructing algebraic equations where excitations are direct or alternating current signals.

(4) The main analysis method for signals and systems is to obtain the responses (output voltages or output currents) of a system to excitations by building differential/difference equations where excitations are periodic or nonperiodic signals.

For instance, in ▶ Figure 4, if u_S and i are, respectively, the excitation and the response of a system or a circuit, how can we obtain the response i of the system under different excitations? From circuit analysis, the currents in ▶ Figure 4a and b are $i = \frac{u_S}{R}$ and $\dot{I} = \frac{\dot{U}_S}{R+j\omega L}$, respectively. Because excitations in ▶ Figure 4c and d are non-sinusoidal periodic and nonperiodic signals, respectively, the currents cannot be obtained. Fortunately, these problems can be solved by Fourier series and Fourier transform in signals and systems.

(5) Circuits Analysis includes solution methods for algebraic equations obtained by laws and theorems of circuits and the phasor analysis method for alternating current circuits.

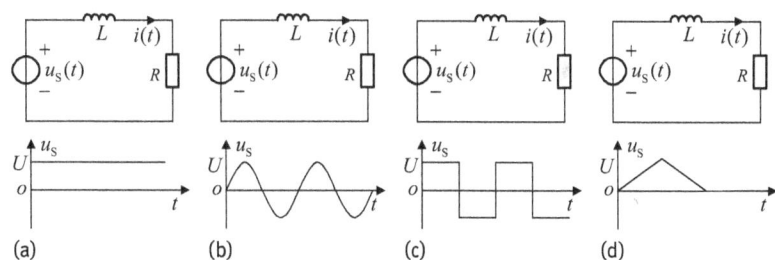

(a)   (b)   (c)   (d)

Fig. 4: Circuits Analysis examples.

1. DC Signals	**Circuits analysis**	1. DC Signals
2. AC Signals	Algebraic equation	2. AC Signals

Excitation $u(t)$ $i(t)$	LTI electric system	Response $u(t)$ $i(t)$
Excitation $u(t)$ $i(t)$	(circuit)	Response $u(t)$ $i(t)$

1. Aeriodic signals	Differential or difference	1. Aeriodic signals
2. Aperiodic signals	equation	2. Aperiodic signals
3. Discrete signals	**Signals and Systems**	3. Discrete signals

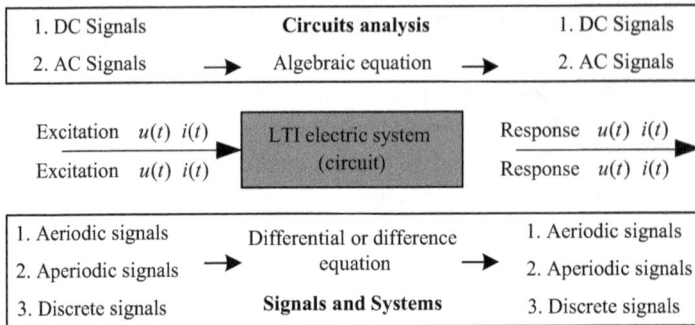

Fig. 5: Main differences between Signals and Systems and Circuits Analysis.

(6) Signals and Systems include solution methods for differential equations in the time, frequency and complex frequency domains and difference equations in the time and z domains.

The main differences and similarities between the two courses are shown in ▸ Figure 5. Note that the points in the figure only focus on electricity technology, in fact, the contents of Signals and Systems can be also used in mechanical systems and other analogous systems.

Status of the course

In conclusion, Signals and Systems is a professional basic course, which takes the signal and the system as the core, has system performance analysis as the purpose and employs mathematics as the tool to establish the mathematical model as the premise and solve the model as the means.

Features of this textbook

The Signals and Systems course is not only a specialized course with professional concepts, but also a mathematical one with massive computations. To help readers master the contents in a better way, we deliberately increased the number of examples (about 140 sets) to expand insights and to improve understanding by analogy. At the same time, we also arranged more than 160 problems with answers to help readers grab and consolidate knowledge learned. In addition, to deepen readers' understanding of the contents and to improve problem solving skills, a section called Solved Questions was added at the end of each chapter, which includes about 50 exam questions and solutions selected from other universities.

Due to space limitations, this book has been split into two volumes. Volume 1 mainly discusses problems concerning continuous-time signals and systems analysis. Volume 2 focuses on issues about discrete time signals and systems analysis.

I would like to express my sincere thanks to Associate Professor Wei-Feng Zhang, the coauthor of the books, and Associate Professors Tao Zhang and Shuai Ren, and lecturers Xiao-Xian Qian and Jian-Fang Xiong for preparing the Solved Questions and translating the Chinese manuscript into English. I also wish to thank Dan-Yun Zheng, Jie-Xu Zhao, Xiang-Yun Li, Pei-Cheng Wang, Juan-Juan Wu, Jing Wu and Run-Qing Li for proofreading all examples, exercises and answers and helping to translate the manuscript. Finally, thanks are due to authors and translators of reference books in the books.

The books are the summary of the authors' teaching experience of many years; any suggestions for improvements to the book would be greatly appreciated. Please feel free to contact us about any problems encountered during reading; we would be grateful for your comments.

June 2016   Weigang Zhang

Contents

Contents of Volume 1

8 Analysis of discrete signals and systems in the time domain

Questions: The analysis methods for continuous signals and systems in the time domain have been discussed. What analysis methods should we use for discrete signals and systems in the time domain?

Solutions: With the help of the analysis methods for continuous signals and systems → solve the difference equations → decompose the responses.

Results: Unit response, convolution sum.

In engineering practice, with the rapid development of computer technology, many problems in continuous time have been increasingly transformed into problems to be processed in discrete time, which has such advantages as high accuracy, stable performance, strong anti-interference ability and flexible handling, etc. Therefore, we should study discrete signals and systems.

Although continuous signals and systems, and discrete signals and systems each have their own strict and effective analysis methods, there are also many corresponding relationships and the similarities between of them in representation, properties and processing procedures. Moreover, these relationships and similarities can help us to seek analysis methods of discrete signals and systems.

In this chapter, we will develop mainly the analysis methods of discrete signals and systems in the time domain.

8.1 Basic discrete signals

There was a preliminary introduction of discrete signals in Chapter 1. In order to analyze discrete systems, first, we need to learn more about discrete signals and their properties.

8.1.1 Periodic sequences

If values of a discrete signal $f[n]$ repeat every N points, this sequence is considered as a periodic sequence with period N and is represented as

$$f[n] = f[n + kN] \quad k = 0, \pm1, \pm2, \ldots \tag{8.1-1}$$

where N is a positive integer.

https://doi.org/10.1515/9783110541205-001

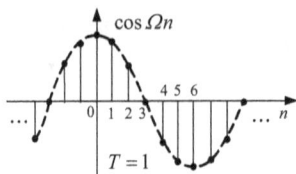

Fig. 8.1: Sinusoidal sequence.

8.1.2 Sinusoidal sequences

A sinusoidal sequence is defined as

$$f[n] \stackrel{\text{def}}{=} \cos \Omega n \quad \text{or} \quad f[n] \stackrel{\text{def}}{=} \sin \Omega n . \tag{8.1-2}$$

This kind of signal can be obtained by the following transformation:

$$\cos \omega t|_{t=nT} = \cos \omega n T|_{\Omega = \omega T} = \cos \Omega n ,$$

where T is the sampling interval but is not the period value of a periodic function like $T_0 = 2\pi/\omega$, and ω is the familiar angular frequency of the continuous wave, $\Omega = \omega T$ is the discrete or digital angular frequency whose unit is "rad". Equation (8.1-2) is illustrated in ▶ Figure 8.1.

It should be noted that $\cos \omega t$ is periodic, but $\cos \Omega n$ is not always a periodic sequence; it will be periodic only if

$$\frac{\Omega}{2\pi} = \frac{m}{N} = \text{a rational number} . \tag{8.1-3}$$

That means that $\cos \Omega n$ can meet the following equation:

$$\cos \Omega n = \cos \Omega (n \pm N) = \cos \Omega (n \pm 2N) = \cos \Omega (n \pm 3N) = \dots$$

In equation (8.1-3), N is the period, and m is an arbitrary integer. For example,
(a) $f[n] = \cos(\pi/6)n$, $(\Omega = \pi/6)$, and $\Omega/2\pi = m/N = 1/12$ is a rational number, so this sequence is a periodic sequence. Its waveform for when $m = 1$, $N = 12$, is shown in ▶ Figure 8.2a.

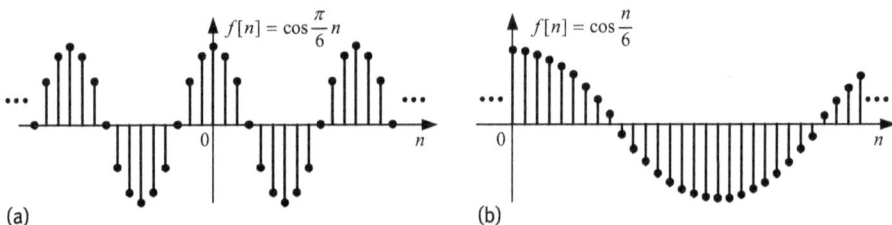

(a) (b)

Fig. 8.2: Periodic sequence and nonperiodic sequence.

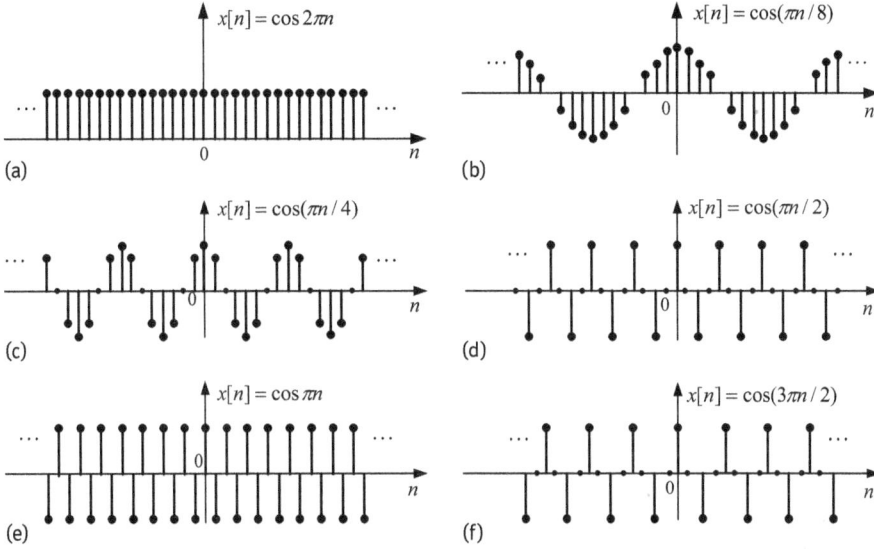

Fig. 8.3: Several sinusoidal sequences with different frequencies.

(b) $f[n] = \cos(1/6)n$, $(\Omega = 1/6)$, and $\Omega/2\pi = m/N = 1/12\pi$ is an irrational number. Because we cannot find an integer m to make N become an integer, the sequence is nonperiodic; it is plotted in ▶ Figure 8.2b.

Note: The envelope of the waveform depicted in ▶ Figure 8.2b is similar to the waveform of a cosine signal, but samples of the sequence do not appear circularly.

To help readers understand the sinusoidal sequence, we show several sinusoidal sequences with different digital frequencies in ▶ Figure 8.3. The waveforms state that when a continuous sinusoidal signal is transformed into discrete form, the discrete waveform may be no longer familiar to us.

8.1.3 Complex exponential sequences

The complex exponential sequence is defined as

$$f[n] \overset{\text{def}}{=} e^{(\rho+j\Omega)n} = e^{\beta n}, \tag{8.1-4}$$

where $\beta = \rho+j\Omega$ is similar to the $s = \sigma+j\omega$ in the continuous time domain. If $\rho = 0$, the sequence will turn into an imaginary exponential sequence, which can be expressed as

$$f[n] = e^{j\omega t}\big|_{t=nT} = e^{j\omega nT}\big|_{\Omega=\omega T} = e^{j\Omega n}. \tag{8.1-5}$$

With Euler's relation, the imaginary exponential sequence can be written as

$$e^{j\Omega n} = \cos \Omega n + j \sin \Omega n . \tag{8.1-6}$$

Similarly to the sinusoidal sequence, if an imaginary exponential sequence is a periodic sequence with N as its period, namely,

$$e^{j\Omega n} = e^{j\Omega(n+N)} , \tag{8.1-7}$$

then it should satisfy the following expression:

$$\frac{\Omega}{2\pi} = \frac{m}{N} = \text{a rational number} . \tag{8.1-8}$$

According to the above contents about sinusoidal and imaginary exponential sequences, the main differences between sequences and the continuous signals are as follows.

(1) A discrete sequence with a periodic envelope is not necessarily periodic.
(2) Periodic sequences with different frequencies are likely to be the same. As we know, all $e^{j\omega t}$ with different ω are different, and the bigger ω, the faster the oscillation of a signal $e^{j\omega t}$. Differently, imaginary exponential sequences with the angular frequency Ω and $(\Omega + 2k\pi)$ are all the same. So, we only need to observe these sequences within a certain range of 2π belonging to Ω. This range is generally arranged as $-\pi \leq \Omega \leq \pi$ or $0 \leq \Omega \leq 2\pi$, namely, the main values range of Ω.

From the analysis, the discrete frequency Ω has two important features:

- When Ω is an integer times 2π, the imaginary exponential sequence is a constant sequence, and sequences whose frequencies are around these frequencies are lower frequency sequences, and these can be similarly considered as continuous signals with frequencies $\omega = 0$ and neighboring frequencies.
- When Ω is an odd multiple of π, the fluctuating frequency of an imaginary exponential sequence be the highest, which corresponds to the case of the continuous angular frequency $\omega \to \infty$ for a continuous signal.

(3) The discrete imaginary exponential or sinusoidal sequence also has the harmonic property, but harmonic waves with different frequencies can be the same (such as those in ▶ Figure 8.3d and f). Among the infinite harmonics of a discrete imaginary exponential sequence, only N of them are periodic sequences different from each other, whereas the others all can find a same sequence from these N sequences.

8.1.4 Exponential sequences

The exponential sequence is defined as

$$f[n] \overset{\text{def}}{=} \begin{cases} 0 & (n < 0) \\ e^{an} & (n \geq 0) \end{cases} , \tag{8.1-9}$$

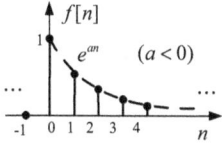

Fig. 8.4: Exponential sequence.

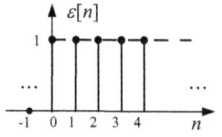

Fig. 8.5: Unit step sequence.

and is sketched in ▶ Figure 8.4, where a is a real number.

8.1.5 Unit step sequence

In the discrete domain, $\varepsilon[n]$ is defined as the unit step sequence,

$$\varepsilon[n] \overset{\text{def}}{=} \begin{cases} 0 & (n < 0) \\ 1 & (n \geq 0) \end{cases}. \tag{8.1-10}$$

Its functions are similar to $\varepsilon(t)$ and the waveform is sketched in ▶ Figure 8.5. Note that $\varepsilon(t)$ is not defined at $t = 0$, but $\varepsilon[n]$ has a certain value 1 at $n = 0$.

8.1.6 Unit impulse sequence

In the discrete domain, $\delta[n]$ is defined as the unit impulse sequence or unit sequence, i.e.

$$\delta[n] \overset{\text{def}}{=} \begin{cases} 0 & (n \neq 0) \\ 1 & (n = 0) \end{cases}. \tag{8.1-11}$$

Its functions are similar to $\delta[n]$, and the waveform is shown in ▶ Figure 8.6. Note that the value of $\delta(t)$ at $t = 0$ is infinite, while the value of $\delta[n]$ is equal to 1 for $n = 0$.

As we know, the continuous signals $\varepsilon(t)$ and $\delta(t)$ are related in a differential relation; then what is the relationship between $\varepsilon[n]$ and $\delta[n]$? Comparing the waveforms

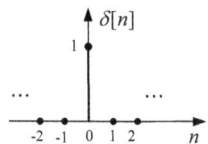

Fig. 8.6: Unit impulse sequence.

Tab. 8.1: Relationships between continuous and discrete singular signals.

Continuous signal	Discrete signal
Differential $\delta(t) = \dfrac{d\varepsilon(t)}{dt}$	Difference $\delta[n] = \varepsilon[n] - \varepsilon[n-1]$
Running integral $\varepsilon(t) = \displaystyle\int_{-\infty}^{t} \delta(\tau)d\tau$	Running sum $\varepsilon[n] = \displaystyle\sum_{k=-\infty}^{n} \delta[k]$

of $\varepsilon[n]$ with $\delta[n]$, their mathematic relations can be obtained by

$$\delta[n] = \varepsilon[n] - \varepsilon[n-1] , \tag{8.1-12}$$

$$\varepsilon[n] = \sum_{m=0}^{\infty} \delta[n-m] \overset{k=n-m}{=} \sum_{k=-\infty}^{n} \delta[k] . \tag{8.1-13}$$

If equations (8.1-12) and (8.1-13) are respectively called a first-order difference equation and a running sum, then they are similar to the differential and the integral operations between $\delta(t)$ and $\varepsilon(t)$ in concept. So, the relationships between equations (8.1-12) and (8.1-13), and $\delta(t)$ and $\varepsilon(t)$ are listed in Table 8.1.

8.1.7 *z* sequence

The z sequence is defined as

$$f[n] \overset{\text{def}}{=} z^n . \tag{8.1-14}$$

In this formula, z is usually a complex number, which is generally expressed as a polar form like $z = |z|\, e^{j\Omega_0} = |z|\, \angle\Omega_0$.

Note: If $z = e^\beta$, then the above equation changes into $f[n] = z^n = e^{\beta n}$, so, the z sequence can be considered as another form of a complex exponential sequence.

Similarly to continuous signals, the complex exponential sequence and the unit impulse sequence are the cores in discrete signals. In addition, like equation (6.8-1), for a discrete LTI system there is also $z^n \rightarrow H(z)z^n$. Obviously, the z sequence is similar to e^{st} in status and role. The details can be seen in Chapter 9.

8.2 Fundamental operations of sequences

8.2.1 Arithmetic operations

The sum, the difference, the product and the quotient of two given sequences are, respectively, a new sequence which is formed by the sums, differences, products and quotients of the corresponding values term by term of two given sequences with the same independent variable n. That is,

$$f[n] = f_1[n] + f_2[n] , \tag{8.2-1}$$
$$f[n] = f_1[n] - f_2[n] , \tag{8.2-2}$$
$$f[n] = f_1[n] \cdot f_2[n] , \tag{8.2-3}$$
$$f[n] = f_1[n]/f_2[n] . \tag{8.2-4}$$

Example 8.2-1. Obtain the sum, the difference, the product, the quotient sequence, respectively, of the two sequences,

$$f_1[n] = \begin{cases} n + 3, & (-2 \leq n \leq 2) \\ 0, & \text{else} \end{cases}, \qquad f_2[n] = \begin{cases} \left(\frac{3}{4}\right)^n + 1 & (0 \leq n \leq 3) \\ 0 & \text{else} \end{cases} .$$

Solution. The sum and difference of $f_1[n]$ and $f_2[n]$ are, respectively,

$$f_1[n] + f_2[n] = \begin{cases} (n+3) + 0 = n + 3, & (-2 \leq n \leq -1) \\ (n+3) + \left[\left(\frac{3}{4}\right)^n + 1\right] = \left(\frac{3}{4}\right)^n + n + 4, & (0 \leq n \leq 2) \\ 0 + \left[\left(\frac{3}{4}\right)^n + 1\right] = \left(\frac{3}{4}\right)^n + 1, & (n = 3) \\ 0 + 0 = 0, & \text{else} \end{cases}$$

$$f_1[n] - f_2[n] = \begin{cases} (n+3) - 0 = n + 3, & (-2 \leq n \leq -1) \\ (n+3) - \left[\left(\frac{3}{4}\right)^n + 1\right] = -\left(\frac{3}{4}\right)^n + n + 2, & (0 \leq n \leq 2) \\ 0 - \left[\left(\frac{3}{4}\right)^n + 1\right] = -\left(\frac{3}{4}\right)^n - 1, & (n = 3) \\ 0 - 0 = 0, & \text{other} \end{cases}$$

The product and quotient of $f_1[n]$ and $f_2[n]$ are, respectively,

$$f_1[n] \cdot f_2[n] = \begin{cases} (n+3)\left(\frac{3}{4}\right)^n + n + 3, & 0 \leq n \leq 2 \\ 0, & \text{other} \end{cases}$$

$$f_1[n]/f_2[n] = \begin{cases} 0, & n = 3 \\ (n+3)\Big/\left[\left(\frac{3}{4}\right)^n + 1\right], & 0 \leq n \leq 2 \\ \text{inexistence}, & \text{other} \end{cases} .$$

The waveforms of $f_1[n], f_2[n]$, and their sum, difference, product are shown in ▶ Figure 8.7.

8.2.2 Time shifting

If the independent variable n of a sequence $f[n]$ is replaced by $n \pm k$ ($k \geq 0$), a new sequence $f[n \pm k]$ is obtained. Obviously, as a time shifting sequence, $f[n + k]$ is generated by left shifting of the original sequence $f[n]$ to k units, while $f[n - k]$ is obtained by right shifting; both are sketched in ▶ Figure 8.8.

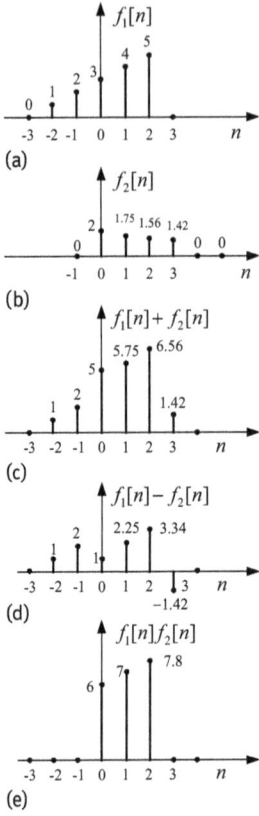

Fig. 8.7: E8.2-1 – Sequences operation scheme.

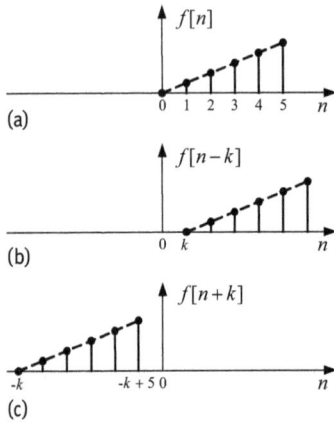

Fig. 8.8: Time shifting scheme of sequence.

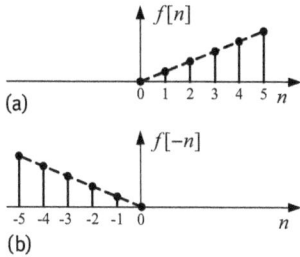

(a)

(b)

Fig. 8.9: Sequence reversal scheme.

8.2.3 Time reversal

If the independent variable n of a signal $f[n]$ is replaced by $-n$, $f[-n]$ is obtained as a new sequence. Obviously, sequences $[n]$ and $f[-n]$ are longitudinal axisymmetric or even symmetric, as shown in ▶ Figure 8.9.

8.2.4 Accumulation

Similarly to the integral of a continuous signal, the accumulation or running sum of a sequence $f[k]$ is defined as

$$y[n] \overset{\text{def}}{=} \sum_{k=-\infty}^{n} f[k] . \tag{8.2-5}$$

Example 8.2-2. Find the accumulation of the unit step sequence.

Solution. With the summation formula in the series, we have

$$\sum_{k=-\infty}^{n} \varepsilon[k] = \left[\sum_{k=0}^{n} 1 \right] \varepsilon[n] = (n+1)\varepsilon[n]$$

8.2.5 Difference

The difference operation of a sequence can be constructed according to the differential operation of a continuous signal. The forward first-order difference of a sequence $f[n]$ is defined as

$$\Delta f[n] \overset{\text{def}}{=} f[n+1] - f[n] . \tag{8.2-6}$$

The backward first-order difference of a sequence $f[n]$ is defined as

$$\nabla f[n] \overset{\text{def}}{=} f[n] - f[n-1] . \tag{8.2-7}$$

For example, the waveforms of a unit step sequence and its time shifting sequence are shown in ▶ Figure 8.10a and b; the first-order backward difference of unit step-sequence is

$$\varepsilon[n] - \varepsilon[n-1] = \delta[n] \tag{8.2-8}$$

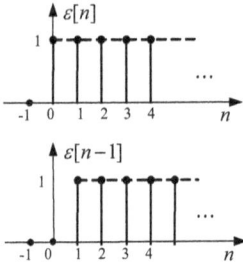

Fig. 8.10: Time shifting scheme of unit step sequence.

Obviously, the unit impulse sequence is the backward first-order difference of unit step sequence.

From the above definitions, it can be seen that the difference operation is very similar to the differential one conceptually. Note that symbols Δ and ∇ represent two different operations. In either the forward or the backward difference operation, the essence of the difference is that two neighboring values of a sequence are subtracted.

Example 8.2-3. Find the differences of following sequences
(1) $y[n] = \sum_{i=0}^{n} f[i]$, find $\Delta y[n]$.
(2) $y[n] = \varepsilon[n]$, find $\nabla y[n-1]$ and $\Delta y[n-1]$.

Solution. (1)

$$y[n] = \sum_{i=0}^{n} f[i] = f[0] + f[1] + \cdots + f[n] ,$$

$$y[n+1] = \sum_{i=0}^{n+1} f[i] = f[0] + f[1] + \cdots + f[n] + f[n+1] ,$$

so,

$$\Delta y[n] = y[n+1] - y[n] = f[n+1] .$$

(2)

$$\nabla y[n-1] = y[n-1] - y[n-2] = \varepsilon[n-1] - \varepsilon[n-2] = \delta[n-1] ,$$
$$\Delta y[n-1] = y[n] - y[n-1] = \varepsilon[n] - \varepsilon[n-1] = \delta[n] .$$

8.2.6 Time scaling

Supposing a sequence is $f[n]$, time scaling can be classified into two types.
(1) If the independent variable n is replaced by an, and a is a positive integer, the relationship between $f[n]$ and $f[an]$ can be represented as

$$f[n] \rightarrow f[an], \quad a = 2, 3, 4, \ldots \tag{8.2-9}$$

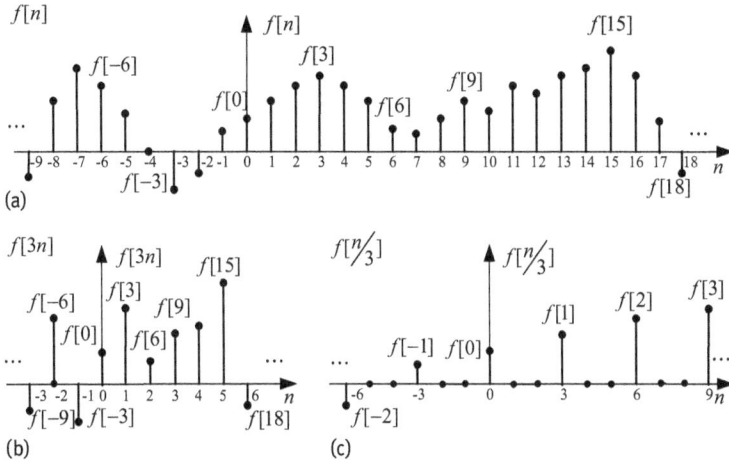

Fig. 8.11: Discrete signal scaling sketch.

(2) If the independent variable n is replaced by $\frac{n}{a}$, and a is a positive integer, the relationship between $f[n]$ and $f[\frac{n}{a}]$ can be represented as

$$f[n] \to \begin{cases} f\left[\frac{n}{a}\right], & n = ka \\ 0, & n \neq ka \end{cases}, \quad k = 0, \pm 1, \pm 2, \ldots \tag{8.2-10}$$

▶ Figure 8.11 shows the waveforms of $f[n]$, $f[3n]$ and $f\left[\frac{n}{3}\right]$, and it is recommended that readers are suggested fully understand the differences between them.

Note: The time scaling of a discrete signal is quite different from that of a continuous signal.
(1) In the discrete time domain, the transform coefficient a can only be a positive integer.
(2) Generally, $f[an]$ or $f\left[\frac{n}{a}\right]$ will not be a signal of which the original signal is compressed or expanded a times in time, and waveforms are totally different from those $f[n]$.

8.2.7 Convolution sum

1. Concept of the convolution sum
Similarly to the definition of the convolution integral, the convolution sum of two discrete sequences $f_1[n]$ and $f_2[n]$ is defined as

$$f_1[n] * f_2[n] \overset{\text{def}}{=} \sum_{k=-\infty}^{\infty} f_1[k]f_2[n-k]. \tag{8.2-11}$$

Usually, the convolution sum can be also written as the discrete convolution.

2. Properties of the convolution sum

(1) Commutative law

$$f_1[n] * f_2[n] = f_2[n] * f_1[n] . \tag{8.2-12}$$

(2) Associative law

$$f_1[n] * [f_2[n] * f_3[n]] = [f_1[n] * f_2[n]] * f_3[n] . \tag{8.2-13}$$

(3) Distributive law

$$f_1[n] * [f_2[n] + f_3[n]] = f_1[n] * f_2[n] + f_1[n] * f_3[n] . \tag{8.2-14}$$

The above laws can be easily proved based on some appropriate variable substitutions and a change of operation orders, which will not be introduced here.

(4) The convolution sum of an arbitrary sequence $f[n]$ and $\delta[n]$ is still the sequence itself,

$$f[n] * \delta[n] = \delta[n] * f[n] = f[n] . \tag{8.2-15}$$

It can be generalized as

$$f[n] * \delta[n - k] = f[n - k] , \tag{8.2-16}$$
$$f[n - k_1] * \delta[n - k_2] = f[n - k_1 - k_2] . \tag{8.2-17}$$

This means that $\delta[n - 1]$ can be regarded as the mathematical model of a unit delayer.

(5) The convolution sum of any sequence $f[n]$ and a unit step sequence is

$$f[n] * \varepsilon[n] = \sum_{i=-\infty}^{n} f[i] , \tag{8.2-18}$$

which shows that $\varepsilon[n]$ can be regarded as the mathematical model of a digital integrator.

It can be generalized as

$$f[n] * \varepsilon[n - k] = \sum_{i=-\infty}^{n-k} f[i] = \sum_{i=-\infty}^{n} f[i - k] . \tag{8.2-19}$$

(6) If $f_1[n] * f_2[n] = f[n]$, then

$$f_1[n - k] * f_2[n] = f_1[n] * f_2[n - k] = f[n - k] . \tag{8.2-20}$$

It can be generalized as

$$f_1[n - k_1] * f_2[n - k_2] = f_1[n - k_2] * f_2[n - k_1] = f[n - k_1 - k_2] . \tag{8.2-21}$$

3. Calculation of the convolution sum

There are usually five ways to calculate the values of a convolution sum, such as with the definition, the graphic method, the direct solution, the table solution and the multiplication solution, which will be introduced in the following.

(1) The definition method

Example 8.2-4. Given sequences $f[n] = \varepsilon[n]$, $h[n] = (0.8)^n \varepsilon[n]$, find the discrete convolution $y[n] = h[n] * f[n]$.

Solution. According to definition of the convolution sum,

$$y[n] = h[n] * f[n] = \sum_{k=-\infty}^{\infty} h[k]f[n-k] = \sum_{k=-\infty}^{\infty} (0.8)^k \varepsilon[k]\varepsilon[n-k]$$

Considering $\varepsilon[k]$ and $\varepsilon[n-k]$, the above equation can be written as

$$y[n] = \left[\sum_{k=0}^{n} (0.8)^k \right] \varepsilon[n] = \frac{1 - 0.8^{n+1}}{1 - 0.8} \varepsilon[n] = 5 \left(1 - 0.8^{n+1} \right) \varepsilon[n] .$$

Note: The function of $\varepsilon[n]$ in the above formula is to limit the domain of $y[n]$ as $n \geq 0$.

According to the example, the solution of discrete convolution is to calculate the sum of a series; two common formulas related to the sum of a series are given in Table 8.2. Table 8.3 shows the discrete convolutions of common sequences.

(2) The graphic method

Similarly to the convolution operation in continuous time, the discrete convolution can also be calculated by using the graphical method according to the following steps:

Step 1: Substituting variables. $f_1[n]$ and $f_2[n]$ are, respectively, replaced by $f_1[k]$ and $f_2[-k]$ whose waveform is actually the mirror image of $f_2[k]$ about the longitudinal axis.

Step 2: Shifting. $f_2[n-k]$ can be obtained by means of $f_2[-k]$ shifted by n units along the horizontal (k) axis. When $n < 0$, $f_2[-k]$ should be shifted to the left by $|n|$ units, while if $n > 0$, it should be shifted to the right by n units.

Step 3: The product of $f_1[k]$ and $f_2[n-k]$ will be calculated point by point, i.e. $f_1[k]f_2[n-k]$.

Tab. 8.2: Common formulas of summation.

No.	Formula		Instruction
1	$\displaystyle\sum_{n=k_1}^{k_2} a^n = $	$\begin{cases} \frac{a^{k_1} - a^{k_2+1}}{1-a} & a \neq 1 \\ k_2 - k_1 + 1 & a = 1 \end{cases}$	k_1, k_2 are integers, and $k_2 > k_1$
2	$\displaystyle\sum_{n=k_1}^{\infty} a^n = \frac{a^{k_1}}{1-a}$	$|a| < 1$	k_1 is an integer

Tab. 8.3: Discrete convolutions of common sequences.

No.	$f_1[n]$	$f_2[n]$	$f_1[n] * f_2[n]$
1	$f[n]$	$\delta[n]$	$f[n]$
2	$f[n]$	$\varepsilon[n]$	$\displaystyle\sum_{i=-\infty}^{n} f[i]$ $\quad n \geq 0$
3	$\varepsilon[n]$	$\varepsilon[n]$	$(n+1)\varepsilon[n]$
4	$n\varepsilon[n]$	$\varepsilon[n]$	$\dfrac{1}{2}(n+1)n\varepsilon[n]$
5	$a^n\varepsilon[n]$	$\varepsilon[n]$	$\dfrac{1-a^{n+1}}{1-a}\varepsilon[n]$
6	$a_1^n\varepsilon[n]$	$a_2^n\varepsilon[n]$	$\dfrac{a_1^{n+1}-a_2^{n+1}}{a_1-a_2}\varepsilon[n],\quad a_1 \neq a_2$
7	$a^n\varepsilon[n]$	$a^n\varepsilon[n]$	$(n+1)a^n\varepsilon[n]$
8	$n\varepsilon[n]$	$a^n\varepsilon[n]$	$\dfrac{n}{1-a}\varepsilon[n] + \dfrac{a(a^n-1)}{(1-a)^2}\varepsilon[n]$
9	$n\varepsilon[n]$	$n\varepsilon[n]$	$\dfrac{1}{6}(n+1)n(n-1)\varepsilon[n]$

Step 4: The values of the discrete convolution at different n can be obtained by summing up the results of Step 3.

The graphical method is presented through the following example.

Example 8.2-5.

$$f_1[n] = \begin{cases} n & 0 \leq n \leq 3 \\ 0 & \text{other} \end{cases} \quad \text{and} \quad f_2[n] = \begin{cases} 1 & 0 \leq n \leq 3 \\ 0 & \text{other} \end{cases}$$

are given, find the convolution sum $y[n] = f_1[n] * f_2[n]$.

Solution. The complete solution procedure is shown in ▶ Figure 8.12. Readers can analyze it for themselves by referring to the above four steps.

(3) The direct method
The direct method is to put variable n into the definition equation of the convolution sum to obtain the resulting sequence.

Example 8.2-6.

$$f[n] = [\underset{\uparrow}{1}, 1, 2] \quad \text{and} \quad h[n] = [\underset{\uparrow}{2}, 2, 3, 3]$$

are given as two sequences. Find their discrete convolution $y[n]$. (Note that numbers above the arrows represent the values of sequences when $n = 0$.)

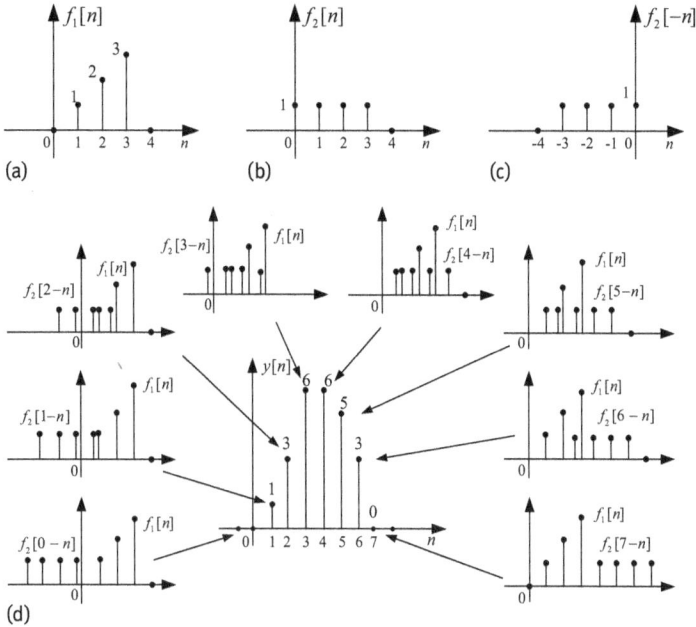

Fig. 8.12: E8.2-5 – Calculating scheme of convolution sum.

Solution. The lengths of the two sequences are limited and start at $n = 0$, so, according to the definition, we have

$$y[n] = f[n] * h[n] = \sum_{k=0}^{n} f[k]h[n-k] .$$

Substituting $n = 0$ into the equation,

$$y[0] = \sum_{k=0}^{0} f[0]h[0-0] = f[0]h[0] = 1 \times 2 = 2 .$$

Substituting $n = 1$ into the equation,

$$y[1] = \sum_{k=0}^{1} f[k]h[n-k] = f[0]h[1] + f[1]h[0] = 2 \times 2 = 4 .$$

Substituting $n = 2$ into the equation,

$$y[2] = \sum_{k=0}^{2} f[k]h[n-k] = f[0]h[2] + f[1]h[1] + f[2]h[0] = 3 + 2 + 4 = 9 .$$

In the same way, we obtain $y[3] = 10$, $y[4] = 9$ and $y[5] = 6$.
Then, the convolution sequence is

$$y[n] = [\underset{\uparrow}{2}, 4, 9, 10, 9, 6] .$$

Point: The length of a convolution sum of two sequences with limited lengths L_1 and L_2 is $L = L_1 + L_2 - 1$.

Obviously, the direct method can easily be calculated by a computer program and can give the convolution sum in sequence form directly, but is difficult to give an analytical formula for it.

(4) The table method
The table method is always used for the convolution sum of the short sequences shown in Table 8.4. In Example 8.2-6, the different values of $f[n]$ and $h[n]$ are, respectively, listed as the first column and the first row, and the product of corresponding elements in each row and each column is regarded as the element in the table. Then, the convolution sum can be obtained by adding elements along the dashed lines.

Tab. 8.4: The table method of Example 8.2-6.

Note that if starting points of $f_1[n]$ and $f_2[n]$ are different, for example $f_1[n]$ starts at $n = 0$ while $f_2[n]$ starts at $n = -k$ ($k > 0$), then we should add k 0 to the left of $f_1[n]$.

(5) The multiplication method
Calculation of the convolution sum of finite length sequences can also done similarly to the ordinary multiplication method. Namely, two sequences are operated according to the rules of multiplication but without carrying, and then the sum of the products in each column is an element of the convolution sequence.

Example 8.2-7. Calculate the convolution sum $y[n]$ of $f_1[n] = [3, 2, 4, 1]$ and $f_2[n] = [2, 1, 5]$.

Solution. Using multiplication rules,

$$
\begin{array}{rrrrrr}
 & 3 & 2 & 4 & 1 & \\
\times & & 2 & 1 & 5 & \\
\hline
 & 15 & 10 & 20 & 5 & \\
 & 3 & 2 & 4 & 1 & \\
+ \quad 6 & 4 & 8 & 2 & & \\
\hline
6 & 7 & 25 & 16 & 21 & 5 \\
\end{array}
$$

we have

$$y[n] = [\underset{\uparrow}{6}, 7, 25, 16, 21, 5] .$$

8.2.8 Sequence energy

The energy of a discrete signal $f[n]$ is defined as

$$E \overset{\text{def}}{=} \sum_{n=-\infty}^{\infty} |f[n]|^2 . \qquad (8.2\text{-}22)$$

8.2.9 Sequence representation with unit impulse sequences

Similarly to the continuous signal $\delta(t)$, the unit impulse sequence $\delta[n]$ can be also used to represent any sequence $f[n]$ with a linear combination form, that is,

$$f[n] = \sum_{k=-\infty}^{+\infty} f[k]\delta[n - k] . \qquad (8.2\text{-}23)$$

8.3 Discrete systems

8.3.1 The concept of discrete systems

If the input and the output of a system are all discrete signals, the system is discrete or digital and can be defined as follows:

A discrete system is the aggregation of related circuits, devices or algorithms, which can change a discrete signal (several) into another discrete signal (others). In other words, a system that can process or transform discrete signals is a discrete system.

As with continuous systems, to analyze discrete systems we must build the mathematical models for them. For the discrete systems in this book, the basic mathematical model in the time domain is a constant coefficient linear difference equation. The discrete system in the time domain is sketched in ▶ Figure 8.13.

8.3.2 Properties of discrete systems

Only linear, time invariant and causal discrete systems will be discussed here, just like the continuous systems. The corresponding conceptions are as follows.

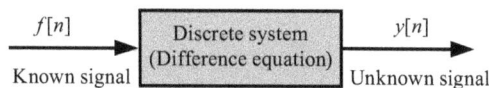

Fig. 8.13: Model of a discrete system in time domain.

1. Response decomposition

If the response resulting from the starting state is $y_x[n]$, and the response produced by the excitation is $y_f[n]$, the complete response $y[n]$ is also

$$y[n] = y_x[n] + y_f[n] . \tag{8.3-1}$$

2. Linearity

If

$$y_1[n] = T[f_1[n]] , y_2[n] = T[f_2[n]] ,$$

then

$$a_1 y_1[n] + a_2 y_2[n] = T[a_1 f_1[n] + a_2 f_2[n]] , \tag{8.3-2}$$

where a_1 and a_2 are arbitrary constants. Note that $f_1[n]$ and $f_2[n]$ can be replaced by the starting states.

3. Time invariance

If

$$y[n] = T[f[n]] ,$$

then

$$y[n - k] = T[f[n - k]] , \tag{8.3-3}$$

where k is a real constant number.

4. Causality

No excitation, no response. The response cannot occur before the excitation appears. In other words, the response at any time only depends on the excitation at the present time and in the past, but is not related to the input in the later, which is the feature of causal systems.

For convenience, a linear time invariant causal discrete system is simply a discrete system in this book without specification. Certainly, all physical systems are causal, whether continuous or discrete.

Example 8.3-1. Determine whether the following systems are linear.
(1) $x[n] \rightarrow y[n] = f^2[n_0] + f^2[n]$;
(2) $x[n] \rightarrow y[n] = \cos\left[an + \frac{\pi}{3}\right] x[n]$

Solution. (1) This system can meet the response decomposition but not the zero-input and zero-state linearities, it is a nonlinear system.
(2) $x_1[n] \rightarrow y_1[n] = \cos\left(an + \frac{\pi}{3}\right) x_1[n]$, $x_2[n] \rightarrow y_2[n] = \cos\left(an + \frac{\pi}{3}\right) x_2[n]$.

Since

$$x_1[n] + x_2[n] \rightarrow y[n] = \cos\left[an + \frac{\pi}{3}\right][x_1[n] + x_2[n]]$$

$$= \cos\left[an + \frac{\pi}{3}\right]x_1[n] + \cos\left[an + \frac{\pi}{3}\right]x_2[n]$$

$$= y_1[n] + y_2[n] ,$$

the system is linear.

Example 8.3-2. Determine whether or not the following systems are time invariant.
(1) $x[n] \rightarrow y[n] = ax[n] + b$;
(2) $x[n] \rightarrow y[n] = nx[n]$

Solution. (1) Because $x[n - n_1] \rightarrow y[n] = ax[n - n_1] + b = y[n - n_1]$, the system is time invariant.
(2) Since $x[n - n_1] \rightarrow y[n] = nx[n - n_1] \neq y[n - n_1] = (n - n_1)x[n - n_1]$, system is time variant.

8.4 Description methods for discrete systems in the time domain

Like for continuous systems, there are main representations based on the difference equation, the operator and the impulse response in the time domain for discrete LTI systems.

8.4.1 Representation with the difference equation

Usually, "difference equation" refers to an equality containing shifted sequences of unknown sequences.

The maximal difference value between left-shifting and right-shifting of an unknown sequence is called the order of the difference equation. The difference equation is very suitable for describing the discrete system, which can reflect directly the input-output relation of a system at discrete time nT (or n).

For example, $y[n]$ represents the population of a country in the nth year, the coefficients a and b are constants, respectively, representing the birth and mortality rates. If $f[n]$ is the net added value of foreign immigrants, the population of this country in $(n + 1)$-th year would be

$$y[n + 1] = y[n] + ay[n] - by[n] + f[n] .$$

Rearranging the equation,

$$y[n + 1] - (1 + a - b)y[n] = f[n] .$$

Obviously, the expression is a difference equation, which is a mathematical model of a population analysis system and can reflect the changes of the population. Because the difference value between the output sequence and its shifted sequence is 1, it is a first-order difference equation. It is not difficult to find that it is also a forward difference equation.

For example, at the beginning of every month a person deposits the money $f[n]$ \$ in a bank. The bank calculates compound interest on a monthly basis (the sum of principal and interest of last month is the principal of this month, the interest rate is α \$/(month.\$)). Then the principal and interest $y[n]$ for the nth month would include three parts, such as the amount of money $f[n]$ deposited for this month, the principal $y[n-1]$ deposited in last month, and the interest $\alpha y[n-1]$ of $y[n-1]$. Therefore, the principal and interest $y[n]$ in the nth month can be described as

$$y[n] = y[n-1] + \alpha y[n-1] + f[n] ,$$

Rearranging the equation,

$$y[n] - (1 + \alpha)y[n-1] = f[n] .$$

This is a first-order backward difference equation.

In summary, we can give the general form of a difference equation now.
(1) The Nth-order forward difference equation is of the form

$$a_N y[n+N] + a_{N-1}y[n+N-1] + \cdots + a_1 y[n+1] + a_0 y[n] =$$
$$b_M f[n+M] + b_{M-1}f[n+M-1] + \cdots + b_1 f[n+1] + b_0 f[n] \quad \text{(8.4-1)}$$

The forward difference equation is more suitable in the state variables analysis method of system.
(2) The Nth-order backward difference equation is of the form

$$a_N y[n] + a_{N-1}y[n-1] + \cdots + a_1 y[n-N+1] + a_0 y[n-N] =$$
$$b_M f[n] + b_{M-1}f[n-1] + \cdots + b_1 f[n-M+1] + b_0 f[n-M] \quad \text{(8.4-2)}$$

The backward difference equation is more likely to analyze the causal system and digital filter, which what is the mainly discussed in this chapter.

Similarly to a differential equation, the significance of a difference equation is the current response $y[n]$ of a discrete system relates not only to the current excitation $f[n]$ but also to past responses such as $y[n-1], y[n-2], \ldots, y[n-k]$. So, a discrete system also has a memory feature.

The above two examples about the difference equation tells us that the contents of this course are suitable not only for the electric systems, but also for other systems (analogous systems) with the difference (differential) equation as the mathematical model.

8.4.2 Representation with the transfer operator

Similarly to the differential operator p in a continuous system, in a discrete system the letter E is defined as the advanced operator, which states an operation of which a sequence is shifted forward (left) one unit; we have $Ey[n] = y[n+1]$, $E^2y[n] = y[n+2]$, $\ldots$, $E^ky[n] = y[n+k]$. Defining $E^{-1} = 1/E$ as the lag operator, which indicates that a sequence is shifted by backward (right) one unit, we have $E^{-1}y[n] = y[n-1]$, $E^{-2}y[n] = y[n-2]$, $\ldots$, $E^{-k}y[n] = y[n-k]$.

Based on these concepts, the forward difference equation (8.4-1) can be expressed as

$$\left(a_N E^N + a_{N-1} E^{N-1} + \cdots + a_1 E + a_0\right) y[n]$$
$$= \left(b_M E^M + b_{M-1} E^{M-1} + \cdots + b_1 E + b_0\right) f[n] . \quad (8.4\text{-}3)$$

The backward difference equation (8.4-2) can be expressed as

$$\left(a_N + a_{N-1} E^{-1} + \cdots + a_1 E^{-(N-1)} + a_0 E^{-N}\right) y[n]$$
$$= \left(b_M + b_{M-1} E^{-1} + \cdots + b_1 E^{-(M-1)} + b_0 E^{-M}\right) f[n] . \quad (8.4\text{-}4)$$

This way, the transfer operator $H(E)$ of a discrete system described by the forward difference equation can be written in the form

$$H(E) = \frac{b_M E^M + b_{M-1} E^{M-1} + \cdots + b_1 E + b_0}{a_N E^N + a_{N-1} E^{N-1} + \cdots + a_1 E + a_0} = \frac{N(E)}{D(E)} , \quad (8.4\text{-}5)$$

where $D(E) = a_N E^N + a_{N-1} E^{N-1} + \cdots + a_1 E + a_0$ is the characteristic polynomial of a discrete system, $D(E) = 0$ is the characteristic equation, and λ_i is also known as characteristic root or natural frequency.

The transfer operator $H(E)$ of a discrete system described by the backward difference equation can be written in the form

$$H(E) = \frac{b_M + b_{M-1} E^{-1} + \cdots + b_1 E^{-(M-1)} + b_0 E^{-M}}{a_N + a_{N-1} E^{-1} + \cdots + a_1 E^{-(N-1)} + a_0 E^{-N}} . \quad (8.4\text{-}6)$$

Obviously, equation (8.4-6) can be rearranged as

$$H(E) = \frac{E^N \left(b_M E^M + b_{M-1} E^{M-1} + \cdots + b_1 E + b_0\right)}{E^M \left(a_N E^N + a_{N-1} E^{N-1} + \cdots + a_1 E + a_0\right)} . \quad (8.4\text{-}7)$$

For causal systems, we have $N \geq M$, so equation (8.4-7) can be rewritten as

$$H(E) = \frac{E^{N-M} \left(b_M E^M + b_{M-1} E^{M-1} + \cdots + b_1 E + b_0\right)}{a_N E^N + a_{N-1} E^{N-1} + \cdots + a_1 E + a_0} . \quad (8.4\text{-}8)$$

Equations (8.4-8) and (8.4-5) have the same denominator, which means the characteristic polynomials of the backward and the forward difference equations are the same.

Equation (8.4-8) states that a backward difference equation can be transformed into a forward difference equation, and a forward difference equation can be transformed into a backward difference equation. This conclusion is similar to the concept that a differential equation and an integral equation can be converted to each other.

For example, the equation

$$y[n] - \frac{5}{6}y[n-1] + \frac{1}{6}y[n-2] = f[n] - \frac{13}{6}f[n-1] + \frac{1}{3}f[n-2]$$

is a second-order backward difference equation. From equation (8.4-6), its transfer operator can be obtained

$$H(E) = \frac{1 - \frac{13}{6}E^{-1} + \frac{1}{3}E^{-2}}{1 - \frac{5}{6}E^{-1} + \frac{1}{6}E^{-2}}$$

From equation (8.4-8), this can be rearranged as

$$H(E) = \frac{E^2 - \frac{13}{6}E + \frac{1}{3}}{E^2 - \frac{5}{6}E + \frac{1}{6}} .$$

Now shifting the original equation forward by two units, we have

$$y[n+2] - \frac{5}{6}y[n+1] + \frac{1}{6}y[n] = f[n+2] - \frac{13}{6}f[n+1] + \frac{1}{3}f[n].$$

According to equation (8.4-5), the transfer operator of the forward difference equation is

$$H(E) = \frac{E^2 - \frac{13}{6}E + \frac{1}{3}}{E^2 - \frac{5}{6}E + \frac{1}{6}} .$$

The transfer operator is the same as the conclusion from equation (8.4-8).

For the difference equations in pure mathematics, the forward and backward equations are just two different forms of a mathematical concept; they can be transformed into each other, as long as we pay attention to the change of the initial conditions. However, in system analysis, the difference equation is a mathematical model of a physical system and has a certain physical meaning. Therefore, we cannot convert at will a backward equation or a forward equation into each other. For instance, a causal system cannot be described by the forward difference equation.

Usually in system analysis, the negative power form of the transfer operator from the backward difference equation (8.4-6) can be arranged to the positive power form [equation (8.4-8)]. As a result, not only discrete system analysis can be unified with the continuous system in the form of the transfer operator $H(p)$, but also system simulation, stability judgment or other research can be expediently achieved (for example, when the corresponding relationship between $h[n]$ and $H(E)$ is discussed, it is convenient to use the positive power form of the $H(E)$). Although $H(E)$ has changed in form,

No.	$H(E)$	$h[n]$ $n \geq 0$
1	1	$\delta[n]$
2	E^m	$\delta[n - m]$
3	$\dfrac{E}{E - \lambda}$	λ^n
4	$\dfrac{E}{E - e^{\gamma T}}$	$e^{\gamma T n}$
5	$\dfrac{E}{(E - \lambda)^2}$	$n\lambda^{n-1}$
6	$\dfrac{E}{(E - e^{\gamma T})^2}$	$n e^{\gamma T(n-1)}$
7	$\dfrac{1}{E - \lambda}$	$\lambda^{n-1}\varepsilon[n - 1]$
8	$\dfrac{1}{E - e^{\gamma T}}$	$e^{\gamma T(n-1)}\varepsilon[n - 1]$
9	$\dfrac{E^2}{(E - \lambda)^2}$	$(n + 1)\lambda^n$

Tab. 8.5: Correspondence relations between $H(E)$ and $h[n]$.

it does not mean that the mathematical model and the characteristics of the system have also changed.

Corresponding to the differential equation of a continuous system $\sum_{i=0}^{n} a_i y^{(i)}(t) = \sum_{j=0}^{m} b_j f^{(j)}(t)$, the forward difference equation of a discrete system can be expressed in the form

$$\sum_{i=0}^{N} a_i y[n + i] = \sum_{j=0}^{M} b_j f[n + j] , \tag{8.4-9}$$

and the backward difference equation of a discrete system can be expressed in the form

$$\sum_{k=0}^{N} a_{N-k} y[n - k] = \sum_{r=0}^{M} b_{M-r} f[n - r] . \tag{8.4-10}$$

Like the continuous system, the discrete system can be also analyzed by using the relation between the transfer operator and the unit impulse response of the system, that is, $h[n]$ can be found by $H(E)$, and then the zero-state response $y_f[n]$ can be obtained by means of the convolution sum method. The correspondence relations between $H(E)$ and $h[n]$ are listed in Table 8.5.

8.4.3 Representation with unit impulse response

Knowing the structure or function of a system is often necessary for a representation with the difference equation or operator. Otherwise, the measuring method should be adopted, in which a specified sequence is input to a system and the output of the system is observed or measured, so the relation between signals on two ports of the system can be obtained. For example, with the premise of the zero starting state, the response of a system to the input $\delta[n]$ is the unit impulse response $h[n]$. As a result, once $h[n]$ has been measured, the zero-state response of the system to any sequence can be represented as $y_f[n] = h[n] * f[n]$. Obviously, these concepts are similar to those of the continuous system.

8.5 Analysis of discrete systems in the time domain

An object processed by a discrete system is a discrete signal. The analysis method for the discrete system is very similar to that for the continuous system, and it is also divided into two categories, such as methods in the time and transform domains. The analysis ideas are also roughly the same, and many concepts correspond to each other, which makes it very convenient for us to analyze discrete systems. Therefore, in the learning process of this chapter, if we are good at introducing corresponding concepts of the continuous system into the analysis of the discrete system, we will obtain twice the result with half the effort. First, this section will discuss the analysis method of discrete system in the time domain.

8.5.1 Classical analysis method

As we know, an important concept in time domain analysis is the state of the system. For a continuous system, its state represents the storage energy situations of all dynamic devices at $t = t_0$, that is, the outputs of the devices when $t = t_0$, and is a set of essential and the fewest data. Similarly, we can define

The state of a discrete system at $n = n_0$ is a set of output values of all the delay components at $n = n_0$. It is also a group of essential and the fewest data such as $x_1[n_0]$, $x_2[n_0]$, ..., $x_n[n_0]$, or simply, $\{x[n_0]\}$.

The response at any time after n_0 can be determined by this group of data and the excitation applying to the system for $n \geq n_0$ together. If we let $n = n_0 = 0$ be the observation moment, then $\{x_1[-1], x_2[-1], x_3[-1], \ldots\}$ will be the starting state values (similar to the moment 0_- in continuous time), and $\{x_1[0], x_2[0], x_3[0], \ldots\}$ will be the initial state values (similar to the moment 0_+ in continuous time).

As with the continuous system, there are also conceptions of the starting and initial conditions for the discrete system.

N starting conditions for an Nth-order discrete system when $n = 0$ are a set of values of $y[n]$ at moments $n = -1, -2, \ldots, -N$, that is, $\{y[-n]\}$, $n = 1, 2, \ldots, N$.

N initial conditions for an Nth-order discrete system when $n = 0$ are a set of values of $y[n]$ at moments $n = 0, 1, 2, \ldots, N - 1$, that is, $\{y[n]\}$, $n = 0, 1, 2, \ldots, N - 1$.

It needs to be explained that as we know that the starting conditions can be converted as the initial conditions by means of the Law of Switching for a continuous system. However, for a discrete system, such a conversion should be achieved by the recursive algorithm, which will not be introduced here in order to save space.

If the starting moment is $n = 0$ **and** $y[-1] = y[-2] = \cdots = y[-N] = 0$ **for an** N**th-order discrete system, this system is a zero-state system.**

1. Iterative analysis method

The analysis of a discrete system is also to solve the system model – the difference equation. As with the solution to the differential equation in continuous time, the difference equation can be also solved in the time and transform domains. The classical analysis method and the response decomposition method can also be applied in the time domain. The classical method is to calculate the homogeneous solution of the equation based on the characteristics of the roots first, then to find a special solution according to the form of the excitation sequence, and finally, to obtain a full or general solution of the equation by adding the two solutions. Response decomposition is to find the zero-state and zero-input responses of the system separately, and then to form the full response by superimposition. In addition, there is also a special iterative method to solve a difference equation, of which the principle is given by the following example.

Example 8.5-1. Suppose that the excitation sequence of discrete system is $f[n] = n\varepsilon[n]$ and the initial condition is $y[0] = 1$. Find the top four terms in the solution for the equation $y[n] + \frac{1}{2}y[n-1] = f[n]$.

Solution. The so-called iterative procedure is to substitute $f[n]$ into the difference equation point by point and to find the corresponding values for $y[n]$.
The original equation can be organized as

$$y[n] = -\frac{1}{2}y[n-1] + f[n] .$$

For $n = 1$,

$$y[1] = -\frac{1}{2}y[0] + f[1] .$$

Substitute $y[0] = 1$ and $f[1] = 1$ into the equation

$$y[1] = -\frac{1}{2} + 1 = 0.5 .$$

For $n = 2$,

$$y[2] = -\frac{1}{2}y[1] + f[2] .$$

Substitute $y[1] = 0.5$ and $f[2] = 2$ into the equation

$$y[2] = -\frac{1}{2} \cdot 0.5 + 2 = 1.75 .$$

For $n = 3$, there is

$$y[3] = -\frac{1}{2} \cdot y[2] + f[3] \, .$$

Substitute $y[2] = 1.75$ and $f[3] = 3$ into the equation

$$y[3] = -\frac{1}{2} \cdot 1.75 + 3 = 2.125$$

Thus,

$$y[n] = [\underset{\uparrow}{1}, 0.5, 1.75, 2.125, \dots] \, .$$

Repeating the above steps, we can obtain the corresponding value of $y[n]$ for any n.

Note that because $y[0]$ is known, the iteration can directly begin from $n = 1$.

Obviously, the calculation of the iterative method is so simple that it is suitable for computer calculation by programming. The disadvantage, however, is that the solution with closed form cannot be easily found. The iterative method is also known as the numerical method.

2. Classical analysis method

In general, the classical analysis method is in accordance with that of the differential equation. Its steps can be summarized as follows.

Step 1: Write out the characteristic equation of the system $D(\lambda) = a_k\lambda^k + a_{k-1}\lambda^{k-1} + \dots + a_1\lambda + a_0 = 0$, and find the characteristic roots.

Step 2: Obtain the homogeneous solution (or free response) $y_c[n]$ according to the characteristic values. If the values are different simple real roots like $\lambda_1, \lambda_2, \dots, \lambda_k$, the homogeneous solution is of the form

$$y_c[n] = C_1\lambda_1^n + C_2\lambda_2^n + \dots + C_k\lambda_k^n = \sum_{i=1}^{k} C_i\lambda_i^n \, . \tag{8.5-1}$$

If the characteristic root λ is a m repeated real root, the homogeneous solution is of the form

$$y_c[n] = (C_1 n^{m-1} + C_2 n^{m-2} + \dots + C_{m-1}n + C_m)\lambda^n = \left[\sum_{i=1}^{m} C_i n^{m-i}\right]\lambda^n \, . \tag{8.5-2}$$

Step 3: Obtain the special solution $y_p[n]$ (or forced response) according to the form of $f[n]$. The commonly used forms of the special solution are listed in Table 8.6.

Step 4: Add the homogeneous and special solutions as the general solution of the difference equation (system's complete response).

$$y[n] = y_c[n] + y_p[n] \, . \tag{8.5-3}$$

Step 5: Substitute the initial values of the system into the above formula and obtain the undetermined coefficients C_i.

Tab. 8.6: Forms of special solutions for the difference equation.

No.	Excitation $f[n]$	Special solution $y_p[n]$
1	A (constant)	C (constant)
2	An	$C_1 n + C_2$
3	An^k	$C_1 n^k + C_2 n^{k-1} + \cdots + C_{k+1}$
4	Ae^{an} (a is real number)	Ce^{an}
5	$Ae^{j\omega n}$	$Ce^{j\omega n}$
6	$A \sin[\omega_0(n + n_0)]$	$C_1 \cos \omega_0 n + C_2 \sin \omega_0 n$
7	$A \cos[\omega_0(n + n_0)]$	$B_1 \cos \omega_0 n + B_2 \sin \omega_0 n$
8	$A\gamma^n$	$C\gamma^n$ (γ is not a characteristic root)
9	$A\gamma^n$	$C_1 n\gamma^n + C_2 \gamma^n$ (γ is a double characteristic root)

Step 6: Substitute the determined coefficients C_i into the general solution.

Note that the constants in Step 7 are written as B_i in order to differentiate them from the constants C_i in Step 6; they are not necessarily the same.

Example 8.5-2. Solve the complete response $y[n]$ of the difference equation $6y[n] - 5y[n-1] + y[n-2] = 10\varepsilon[n]$ using the classical method. $y[0] = 15$ and $y[1] = 9$ are known.

Solution. The characteristic equation of the system is

$$6\lambda^2 - 5\lambda + 1 = 0 .$$

The characteristic roots are

$$\lambda_1 = \frac{1}{2} \quad \text{and} \quad \lambda_2 = \frac{1}{3} .$$

The natural response is

$$y_c[n] = C_1 \left(\frac{1}{2}\right)^n + C_2 \left(\frac{1}{3}\right)^n .$$

The excitation is $f[n] = 10\varepsilon[n]$, so the special solution can be $y_p[n] = C$. Obviously,

$$y_p[n-1] = y_p[n-2] = C .$$

Substituting this into the original equation, we obtain

$$6C - 5C + C = 10 ,$$

and

$$C = 5 .$$

So, the special solution is

$$y_p[n] = 5 .$$

The complete response is

$$y[n] = y_c[n] + y_p[n] = C_1 \left(\frac{1}{2}\right)^n + C_2 \left(\frac{1}{3}\right)^n + 5 .$$

Substituting the initial conditions into this, we have

$$y[0] = C_1 + C_2 + 5 = 15$$

$$y[1] = \frac{1}{2}C_1 + \frac{1}{3}C_2 + 5 = 9 ,$$

and then

$$C_1 = 4, C_2 = 6$$

The complete response is

$$y[n] = y_c[n] + y_p[n] = 4 \left(\frac{1}{2}\right)^n + 6 \left(\frac{1}{3}\right)^n + 5 \quad n \geq 0 .$$

Sometimes the classical method is simpler than the response decomposition method, which will be introduced later, but it has two shortcomings. First, it is difficult to distinguish the zero-input and zero-state responses in the final result. Second, if the form of excitation $f[n]$ is more complex, it is not easy to determine the form of the special solution.

8.5.2 Unit impulse response

Before the introduction of the response decomposition method, we will discuss the unit impulse response of a discrete system. In a continuous system, we know that the unit impulse response $h(t)$ can be used not only to characterize the inherent properties of the system itself, but also to obtain the zero-state response $y_f(t)$ of the system by the convolution with an excitation $f(t)$. In the analysis of discrete systems, we can also define a unit impulse response $h[n]$, whose the effects and characteristics are similar to those of $h(t)$.

The zero-state response of system to the unit sequence $\delta[n]$ is called the unit impulse response or, simply, the unit response, denoted as $h[n]$.

From the definition of $\delta[n]$, we have is $\delta[n] = 0$ for $n > 0$, which means when $n > 0$, $h[n] = 0$. It is found that the unit response $h[n]$ and the homogeneous solution of the difference equation of the system should be the same in the form. The response of the system for $n = 0$ can be considered as the initial condition of $h[n]$. The methods to solve the unit response $h[n]$ in the time domain are given in the following examples.

Example 8.5-3. Solve the unit response of a first-order causal system expressed as

$$y[n] + 0.5y[n - 1] = f[n] .$$

Solution. The characteristic equation is

$$\lambda + 0.5 = 0 .$$

The characteristic root is

$$\lambda = -0.5 .$$

Then, the unit response is

$$h[n] = c(-0.5)^n \varepsilon[n] . \tag{8.5-4}$$

According to the definition of the unit response and the system function, we have

$$h[n] + 0.5h[n-1] = \delta[n] . \tag{8.5-5}$$

Substituting $n = 0$ into expression (8.5-5), we have

$$h[0] + 0.5h[-1] = \delta[0] . \tag{8.5-6}$$

As a causal system, $h[-1] = 0$ and $\delta[0] = 1$, equation (8.5-6) can be changed into

$$h[0] = 1 . \tag{8.5-7}$$

If $n = 0$, equation (8.5-4) can be changed into

$$h[0] = c(-0.5)^0 \varepsilon[0] ,$$

so,

$$c = 1 .$$

Then the unit response of this system is

$$h[n] = (-0.5)^n \varepsilon[n] .$$

Example 8.5-4. Find the unit response $h[n]$ of a second-order system expressed as

$$y[n] + \frac{1}{6}y[n-1] - \frac{1}{6}y[n-2] = f[n] .$$

Solution. The characteristic equation is

$$\lambda^2 + \frac{1}{6}\lambda - \frac{1}{6} = 0 .$$

The characteristic roots are

$$\lambda_1 = \frac{1}{3} \quad \text{and} \quad \lambda_2 = -\frac{1}{2}.$$

So, the unit response is

$$h[n] = \left[c_1 \left(\frac{1}{3}\right)^n + c_2 \left(-\frac{1}{2}\right)^n \right] \varepsilon[n] . \tag{8.5-8}$$

According to the definitions of the unit response and the system function, we have

$$h[n] + \frac{1}{6}h[n-1] - \frac{1}{6}h[n-2] = \delta[n] .$$

As a causal system, $h[-1] = h[-2] = 0$ and $\delta[0] = 1$, so

$$h[0] = -\frac{1}{6}h[-1] + \frac{1}{6}h[-2] + \delta[0] = 1$$

$$h[1] = -\frac{1}{6}h[0] + \frac{1}{6}h[-1] + \delta[1] = -\frac{1}{6}$$

Substituting $h[0] = 1$ and $h[1] = -\frac{1}{6}$ into equation (8.5-8), the simultaneous equations can be obtained by

$$\begin{cases} c_1 + c_2 = 1 \\ \frac{1}{3}c_1 - \frac{1}{2}c_2 = -\frac{1}{6} \end{cases}$$

So,

$$c_1 = \frac{2}{5}, c_2 = \frac{3}{5}$$

The unit response of the system is

$$h[n] = \left[\frac{2}{5}\left(\frac{1}{3}\right)^n + \frac{3}{5}\left(-\frac{1}{2}\right)^n \right] \varepsilon[n] .$$

Example 8.5-5. Solve unit response $h[n]$ of a second-order system with the equation

$$y[n] + \frac{1}{6}y[n-1] - \frac{1}{6}y[n-2] = f[n] - 2f[n-2] .$$

Solution. The response of the system can be seen as the sum of two responses produced by $f[n]$ and $-2f[n-2]$ alone. According definition of the unit response,

$$h[n] = h_1[n] - 2h_1[n-2] ,$$

where $h_1[n]$ is the zero-state response to the excitation $\delta[n]$.

According to Example 8.5-4

$$h_1[n] = \left[\frac{2}{5}\left(\frac{1}{3}\right)^n + \frac{3}{5}\left(-\frac{1}{2}\right)^n \right] \varepsilon[n] .$$

So,

$$-2h_1[n-2] = -2\left[\frac{2}{5}\left(\frac{1}{3}\right)^{n-2} + \frac{3}{5}\left(-\frac{1}{2}\right)^{n-2} \right] \varepsilon[n-2] .$$

Hence, the unit response of the system is

$$h[n] = \left[\frac{2}{5}\left(\frac{1}{3}\right)^n + \frac{3}{5}\left(-\frac{1}{2}\right)^n \right] \varepsilon[n] - 2\left[\frac{2}{5}\left(\frac{1}{3}\right)^{n-2} + \frac{3}{5}\left(-\frac{1}{2}\right)^{n-2} \right] \varepsilon[n-2] .$$

Besides the above methods, the unit response $h[n]$ can also be obtained by the inverse transform of the system function $H(z)$. This is similar to getting $h(t)$ from the inverse transform of the system function $H(s)$ in continuous system analysis and will be discussed later.

8.5.3 Unit step response

Similarly as for the continuous system, the unit step response of the discrete system can be defined as:

The zero-state response of a discrete system to a unit step sequence $\varepsilon[n]$ is called the unit step response, which is represented as $g[n]$.

The relation between the unit step response $g[n]$ and the unit impulse response $h[n]$ is

$$g[n] = \sum_{k=-\infty}^{n} h[k] \qquad (8.5\text{-}9)$$

or

$$h[n] = g[n] - g[n-1] . \qquad (8.5\text{-}10)$$

8.5.4 Analysis with response decomposition

Similarly to the continuous system, the response $y[n]$ of the discrete system can be divided into two parts, such as the zero-input response $y_x[n]$ and the zero-state response $y_f[n]$,

$$y(n) = y_x[n] + y_f[n] . \qquad (8.5\text{-}11)$$

The zero-input response $y_x[n]$ can be described as the response caused only by the starting state (starting condition) of the system when the input sequences are zero.

The zero-state response $y_f[n]$ can be described as the response caused only by the inputs (excitations) of the system when the starting states of the system are zero.

Note: The number of starting states in a system is equal to the order of the system, which is similar to the case of the continuous system.

From the above concept, the response decomposition analysis method is a procedure where the zero-input response $y_x[n]$ and the zero-state response $y_f[n]$ are, respectively, solved and further added together to form the complete response of the system.

Next, the analysis process of this method will be shown via examples.

1. Solving method for the zero-input response

From the definition of the zero-input response, it is the same as the homogeneous solution of the difference equation in form. The coefficients in the homogeneous solution are determined by the starting conditions that are independent of the excitations.

Example 8.5-6. The difference equation of a discrete system is $3y[n] + 2y[n-1] = f[n]$, and we know $y_x[-1] = 0.5$. Solve the zero-input response of the system.

Solution. The characteristic equation is

$$3\lambda + 2 = 0 .$$

The characteristic root is

$$\lambda = -2/3 .$$

So, the zero-input response is

$$y_x[n] = c_x \left(-\frac{2}{3}\right)^n .$$

Substituting the starting condition $y_x[-1] = 0.5$ into the expression above, yields

$$y_x[-1] = c_x \left(-\frac{2}{3}\right)^{-1} = 0.5 ,$$

so,

$$c_x = -\frac{1}{3} .$$

Therefore, the zero-input response is

$$y_x[n] = -\frac{1}{3} \left(-\frac{2}{3}\right)^n , \quad n \geq -1 .$$

Example 8.5-7. A second-order system model is $y[n] + 3y[n-1] + 2y[n-2] = f[n]$, $f[n] = 2^n \varepsilon[n]$, and the initial conditions are $y[0] = 0$, $y[1] = 2$. Solve the zero-input response.

Solution. The system characteristic equation is

$$\lambda^2 + 3\lambda + 2 = 0 .$$

The characteristic roots are

$$\lambda_1 = -1 \quad \text{and} \quad \lambda_2 = -2.$$

The zero-input response is

$$y_x[n] = c_1(-1)^n + c_2(-2)^n . \tag{8.5-12}$$

The undetermined coefficients should be determined by the starting conditions rather than the initial conditions $y[0] = 0$ and $y[1] = 2$. Because the excitation $f[n]$ acts on the system at $n = 0$, the starting conditions are $y[-2]$ and $y[-1]$. Since $y[-2] = y_x[-2]$ and $y[-1] = y_x[-1]$, so $y[-2]$ and $y[-1]$ are found by the iterative method as follows.
Substituting $n = 1$ into the original difference equation, we have

$$y[1] + 3y[0] + 2y[-1] = f[1] ,$$

and we obtain

$$y[-1] = 0 .$$

Substituting $n = 0$ into the original difference equation, we have

$$y[0] + 3y[-1] + 2y[-2] = f[0]$$

and

$$y[-2] = \frac{1}{2} \, .$$

Substituting the starting conditions $y[-1] = 0$ and $y[-2] = \frac{1}{2}$ into equation (8.5-12), we obtain the simultaneous equations

$$\begin{cases} y[-1] = -c_1 - \frac{1}{2}c_2 = 0 \\ y[-2] = c_1 + \frac{1}{4}c_2 = \frac{1}{2} \end{cases} \, .$$

So,

$$c_1 = 1, \quad c_2 = -2 \, .$$

Hence, the zero-input response is

$$y_x[n] = (-1)^n - 2(-2)^n, \quad n \geq -2 \, .$$

2. Solution method for zero-state response

We first find the relation between the zero-state and the unit responses of a discrete system. Assuming the system's excitation is $f[n]$ and the zero-state response is $y_f[n]$ (as illustrated in ▶ Figure 8.14), the steps to solve the question are as follows.

(1) Excitation produces response: $f[n] \rightarrow y_f[n]$.
(2) Definition of the unit response: $\delta[n] \rightarrow h[n]$.
(3) Time invariant feature: $\delta[n - k] \rightarrow h[n - k]$.
(4) Homogeneity: $f[k]\delta[n - k] \rightarrow f[k]h[n - k]$.
(5) Additivity: $\sum_{k=-\infty}^{\infty} f[k]\delta[n - k] \rightarrow \sum_{k=-\infty}^{\infty} f[k]h[n - k]$.
(6) Definition of the convolution sum: $f[n] * \delta[n] \rightarrow f[n] * h[n]$.
(7) Since $f[n] * \delta[n] = f[n]$,

$$f[n] \rightarrow f[n] * h[n] \, .$$

This yields

$$y_f[n] = f[n] * h[n] = \sum_{k=0}^{n} f[k]h[n - k] \, . \tag{8.5-13}$$

Equation (8.5-13) is the expected result.

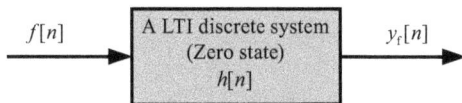

Fig. 8.14: Zero-state response scheme.

The zero-state response of a discrete system is equal to the convolution sum of the excitation sequence and the unit response.

Example 8.5-8. A second-order model is $y[n] - \frac{5}{6}y[n-1] + \frac{1}{6}y[n-2] = f[n]$. Solve the zero-state response when $f[n] = \varepsilon[n]$.

Solution. The system characteristic equation is

$$\lambda^2 - \frac{5}{6}\lambda + \frac{1}{6} = 0 .$$

The characteristic roots are

$$\lambda_1 = \frac{1}{2} \quad \text{and} \quad \lambda_2 = \frac{1}{3}.$$

The unit response $h(n)$ is

$$h[n] = \left[c_1 \left(\frac{1}{2}\right)^n + c_2 \left(\frac{1}{3}\right)^n \right] \varepsilon[n] . \tag{8.5-14}$$

According to the definitions of the unit response and system equation, we have

$$h[n] - \frac{5}{6}h[n-1] + \frac{1}{6}h[n-2] = \delta[n] .$$

Substituting $n = 0$ and $n = 1$ into the expression, we obtain

$$h[0] - \frac{5}{6}h[-1] + \frac{1}{6}h[-2] = \delta[0]$$

$$h[1] - \frac{5}{6}h[0] + \frac{1}{6}h[-1] = \delta[1] .$$

For a causal system, we have

$$h[-1] = h[-2] = 0, \quad \delta[0] = 1 \quad \text{and} \quad \delta[1] = 0.$$

So,

$$h[0] = 1 \quad \text{and} \quad h[1] = \frac{5}{6}.$$

Substituting these initial values into equation (8.5-14), we obtain the simultaneous equations

$$\begin{cases} c_1 + c_2 = 1 \\ \frac{1}{2}c_1 + \frac{1}{3}c_2 = \frac{5}{6} \end{cases} .$$

So,

$$c_1 = 3, \quad c_2 = -2 .$$

Then, the unit response $h[n]$ is

$$h[n] = \left[3 \left(\frac{1}{2}\right)^n - 2 \left(\frac{1}{3}\right)^n \right] \varepsilon[n] .$$

The zero-state response is

$$y_f[n] = f[n] * h[n] = \sum_{k=-\infty}^{\infty} \varepsilon[k]\left[3\left(\frac{1}{2}\right)^{n-k} - 2\left(\frac{1}{3}\right)^{n-k}\right]\varepsilon[n-k]$$

$$= \left\{\sum_{k=0}^{n}\left[3\left(\frac{1}{2}\right)^{n-k} - 2\left(\frac{1}{3}\right)^{n-k}\right]\right\}\varepsilon[n]$$

$$= \left[3\left(\frac{1}{2}\right)^{n}\frac{1-2^{n+1}}{1-2} - 2\left(\frac{1}{3}\right)^{n}\frac{1-3^{n+1}}{1-3}\right]\varepsilon[n]$$

$$= \left[3 - 3\left(\frac{1}{2}\right)^{n} + \left(\frac{1}{3}\right)^{n}\right]\varepsilon[n].$$

Note: The unit step sequence $\varepsilon[n]$ at the right of the zero-state response cannot be omitted, because it represents the variable n beginning from 0. If it is omitted, $n \geq 0$ is needed to mark after the zero-state response. This concept is similar to that of the continuous system.

Based on the above solution methods for the zero-input and the zero-state responses, we obtain the complete response of a system in the time domain. The solution method for the complete response is given by the following examples.

Example 8.5-9. The model of a system is $y[n] - 3y[n-1] + 2y[n-2] = f[n] + f[n-1]$, the starting conditions are $y_x[-2] = 3$ and $y_x[-1] = 2$.
(1) Solve the zero-input response $y_x[n]$.
(2) Solve the unit response $h[n]$ and the step response $g[n]$.
(3) If $f[n] = 2^n\varepsilon[n]$, solve the zero-state response and the complete response.

Solution. (1) The characteristic equation of the system is

$$\lambda^2 - 3\lambda + 2 = 0.$$

The characteristic roots are

$$\lambda_1 = 1, \quad \lambda_2 = 2.$$

The zero-input response is

$$y_x[n] = c_1 + c_2 2^n.$$

Substituting $y_x[-2] = 3$ and $y_x[-1] = 2$ into the expression, we obtain the simultaneous equations

$$\begin{cases} c_1 + \frac{1}{4}c_2 = 3 \\ c_1 + \frac{1}{2}c_2 = 2 \end{cases}.$$

so,

$$c_1 = 4, \quad c_2 = -4.$$

Therefore, the zero-input response of the system is

$$y_x[n] = 4\left(1 - 2^n\right), \quad n \geq -2.$$

(2) The unit response can be obtained by a method similar to Example 8.5-5, but the method of the transfer operator expressed in a partial fraction will be used here. The transfer operator is

$$H(E) = \frac{1 + E^{-1}}{1 - 3E^{-1} + 2E^{-2}} = \frac{E^2 + E}{E^2 - 3E + 2}$$

$$= E\frac{E + 1}{E^2 - 3E + 2} = E\left(\frac{3}{E - 2} - \frac{2}{E - 1}\right) = \frac{3E}{E - 2} - \frac{2E}{E - 1}.$$

According to Table 8.5, the unit response of the system is

$$h[n] = (3 \cdot 2^n - 2)\, \varepsilon[n]\, .$$

The step response $g[n]$ is

$$g[n] = \varepsilon[n] * h[n] = \sum_{k=-\infty}^{\infty} h[k]\varepsilon[n - k] = \left[\sum_{k=0}^{n} h[k]\right]\varepsilon[n]\, .$$

Substituting $h[n] = (3 \cdot 2^n - 2)$ into the equation, yields

$$g[n] = \left[\sum_{k=0}^{n} \left(3 \cdot 2^k - 2\right)\right]\varepsilon[n]$$

$$= [-2(n + 1) + 3\left(2^{n+1} - 1\right)]\varepsilon[n]$$

$$= \left(3 \cdot 2^{n+1} - 2n - 5\right)\varepsilon[n]\, .$$

(3) The zero-state response of the system is

$$y_f[n] = f[n] * h[n] = \left[\sum_{k=0}^{n} 2^k \left(3 \cdot 2^{n-k} - 2\right)\right]\varepsilon[n] = [(3n - 1)\, 2^n + 2]\, \varepsilon[n]\, .$$

The complete response is

$$y[n] = y_x[n] + y_f[n] = 4\left(1 - 2^n\right) + (3n - 1)2^n + 2 = (3n - 5)2^n + 6, \quad n \geq 0\, .$$

Example 8.5-10. $y[n] - 0.7y[n - 1] + 0.1y[n - 2] = 7\varepsilon[n] - 2\varepsilon[n - 1]$ is a difference equation of system, and the initial values of the complete response are $y[0] = 9$ and $y[1] = 13.9$. Solve the zero-input response $y_x[n]$ and the zero-state response $y_f[n]$ of the system.

Solution. The characteristic equation is

$$\lambda^2 - 0.7\lambda + 0.1 = 0\, .$$

The characteristic roots are

$$\lambda_1 = 0.5, \quad \lambda_2 = 0.2\, .$$

The zero-input response is

$$y_x[n] = c_1(0.5)^n + c_2(0.2)^n .$$ (8.5-15)

The undetermined coefficients can be determined by the starting conditions. Considering excitation $f[n]$ acting on the system when $n = 0$, the starting conditions are $y[-2]$ and $y[-1]$.

Substituting $n = 1$ into the original equation, we have

$$y[1] - 0.7y[0] + 0.1y[-1] = 7\varepsilon[1] - 2\varepsilon[0] ,$$

so,

$$y[-1] = -26 .$$

Substituting $n = 0$ into the original equation, we have

$$y[0] - 0.7y[-1] + 0.1y[-2] = 7\varepsilon[0] - 2\varepsilon[-1] ,$$

so,

$$y[-2] = -202 .$$

Substituting the starting conditions $y[-1] = -26$ and $y[-2] = -202$ into equation (8.5-15), we obtain the simultaneous equations

$$\begin{cases} y[-1] = c_1(0.5)^{-1} + c_2(0.2)^{-1} = -26 \\ y[-2] = c_1(0.5)^{-2} + c_2(0.2)^{-2} = -202 \end{cases} .$$

so,

$$c_1 = 12, \quad c_2 = -10 .$$

Therefore, the zero-input response is

$$y_x[n] = 12(0.5)^n - 10(0.2)^n, \quad n \geq -2 .$$

According to the difference equation, the transfer operator is obtained

$$H(E) = \frac{7 - 2E^{-1}}{1 - 0.7E^{-1} + 0.1E^{-2}} = \frac{7E^2 - 2E}{E^2 - 0.7E + 0.1} = E\frac{5}{E - 0.5} + E\frac{2}{E - 0.2} .$$

Therefore, the unit response is

$$h[n] = [5(0.5)^n + 2(0.2)^n] \varepsilon[n] .$$

The zero-state response is

$$\begin{aligned} y_f[n] &= f[n] * h[n] \\ &= \varepsilon[n] * [5(0.5)^n + 2(0.2)^n] \varepsilon[n] \\ &= \varepsilon[n] * 5(0.5)^n \varepsilon[n] + \varepsilon[n] * 2(0.2)^n \varepsilon[n] . \end{aligned}$$

According to Table 8.3, we have

$$y_f[n] = [12.5 - 5(0.5)^n - 0.5(0.2)^n]\, \varepsilon[n]\,.$$

The complete response is

$$y[n] = y_x[n] + y_f[n] = 12(0.5)^n - 10(0.2)^n + 12.5 - 5(0.5)^n - 0.5(0.2)^n$$
$$= 7(0.5)^n - 10.5(0.2)^n + 12.5,\quad n \geq 0\,.$$

Similarly to the continuous system, the response of the discrete system can also be divided into natural response and forced response, transient response and steady state response.

(1) The homogeneous solution of the difference equation is the natural response, and the special solution is the forced response.

(2) The natural response corresponding to the characteristic root with $|\lambda| < 1$ is also the transient response. For almost all physical systems, the natural response is also the transient response.

(3) Whole zero-input response and a part in the zero-state response form the natural response; the remaining component in the zero-state response is the forced response.

(4) The steady state response comes from the forced response. The forced response can also contain the transient response (components).

For example, in the $y[n]$ of Example 8.5-10, the terms $7(0.5)^n - 10.5(0.2)^n$ are both the natural and transient responses, and term 12.5 is both the forced response and steady state response.

8.6 Testing for memorability, invertibility and causality

Similarly to the continuous system, if and only if the unit response $h[n]$ satisfies

$$h[n] = K\delta[n]\,, \tag{8.6-1}$$

then the response and excitation of a system are related by

$$y[n] = Kf[n]\,. \tag{8.6-2}$$

As a result, the discrete system is a memoryless (static) system. In other words, **a system described by the difference equation is a memory system.**

If the unit response of a discrete system is $h[n]$ and another system's is $h_i[n]$, when

$$h[n] * h_i[n] = \delta[n]\,, \tag{8.6-3}$$

then the system with $h[n]$ as the unit response is called a reversible or an original system, and the system with $h_i[n]$ as the unit response is called an inverse system of the original system.

If and only if the unit response $h[n]$ of a discrete system satisfies

$$h[n] = 0, \quad n < 0, \tag{8.6-4}$$

then this is a causal system. Or we can say that a system with a causal sequence as its unit response is causal.

Example 8.6-1. Judge the causality of the following systems.
(1) $h[n] = \varepsilon[3 - n]$;　　　　(2) $h[n] = \delta[n + 4]$;　　　　(3) $h[n] = 0.5^n \varepsilon[n]$

Solution. (1) When $n = -1$, $h[n] = \varepsilon[3 - n] = \varepsilon[4] = 1 \neq 0$, so it is not a causal system.
(2) When $n = -4$, $h[n] = \delta[n + 4] = \delta[0] = 1 \neq 0$, so it is not a causal system.
(3) When $n < 0$, $h[n] = 0$, so it is a causal system.

8.7 Solved questions

Question 8-1. For the excitation and the unit response of a system shown in ▸ Figure Q8-1, find the zero-state response $y_f[n]$ using the definition of the convolution sum.

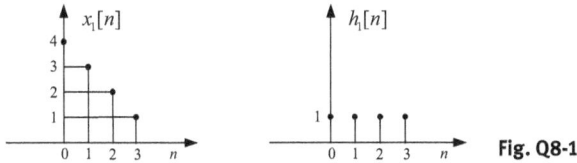

Fig. Q8-1

Solution. From ▸ Figure Q8-1, we know that

$$x_1[n] = 4\delta[n] + 3\delta[n - 1] + 2\delta[n - 2] + \delta[n - 3],$$
$$h_1[n] = \delta[n] + \delta[n - 1] + \delta[n - 2] + \delta[n - 3].$$

According to the convolution theorem, we obtain the zero-state response as

$$y_f[n] = x_1[n] * h_1[n] = \{4, 7, 9, 10, 6, 3, 1\}.$$

Question 8-2. If $x[n] = \{0, \underset{\uparrow}{1}, 2, 3, 4, 3, 2, 1\}$ is known, then $x[2n] =$ _____

Solution. According to time scaling, the answer is $x[2n] = \{\underset{\uparrow}{0}, 2, 4, 2\}$.

Question 8-3. $x_1[n] = \{1, \underset{\uparrow}{3}, -1, 0, 0\}$ and $x_2[n] = \{\underset{\uparrow}{3}, 1, 0, 0, 2\}$ are known. Find the discrete convolution $x[n] = x_1[n] * x_2[n]$.

Solution. According to the table method, we have

n $x_1(n)$ $n_2(n)$	-1	0	1	2	3	4
	1	3	-1	0	0	0
-1 0	0	0	0	0	0	0
0 3	3	9	-3	0	0	0
1 1	1	3	-1	0	0	0
2 0	0	0	0	0	0	0
3 0	0	0	0	0	0	0
4 2	2	6	-2	0	0	0

Moreover,

$$x[n] = \{3, 10, 0, -1, 2, 6, -2\} .$$

According to the multiplication method, we have

```
                    3    1    0    0    2
           ×                  1    3   -1
          ─────────────────────────────────
                   -3   -1    0    0   -2
               9    3    0    0    2
        +  3    1    0    0    2
          ─────────────────────────────────
           3   10    0   -1    2    6   -2
```

So, the discrete convolution is

$$x[n] = \{3, 10, 0, -1, 2, 6, -2\} .$$

Question 8-4. Using the recursive algorithm, find the unit response $h[n]$ of the causal system described by the following difference equation and calculate the first four terms at least.

$$y[n] + \frac{1}{2}y[n-1] - \frac{1}{2}y[n-2] = \sum_{k=0}^{\infty} f[n-k] .$$

Solution. From the equation, the unit impulse response should meet

$$h[n] + \frac{1}{2}h[n-1] - \frac{1}{2}h[n-2] = \sum_{k=0}^{\infty} \delta[n-k] = \varepsilon[n] .$$

Rearranging the equation above, we obtain

$$h[n] = \varepsilon[n] - \frac{1}{2}h[n-1] + \frac{1}{2}h[n-2] .$$

Since the system is causal, $h[n] < 0$ for $n < 0$. Setting $n = 0, 1, 2, 3, \ldots$ separately,

$$h[0] = \varepsilon[0] - \frac{1}{2}h[-1] + \frac{1}{2}h[-2] = 1, \quad h[1] = \varepsilon[1] - \frac{1}{2}h[0] + \frac{1}{2}h[-1] = \frac{1}{2},$$

$$h[2] = \varepsilon[2] - \frac{1}{2}h[1] + \frac{1}{2}h[0] = \frac{5}{4}, \quad h[3] = \varepsilon[3] - \frac{1}{2}h[2] + \frac{1}{2}h[1] = \frac{5}{8}.$$

Question 8-5. The input of a system S is $f[n]$, and the output is $y[n]$. This system consists of S_1 and S_2 in cascade form, and the input-output relationships of S_1 and S_2 are separately $y_1[n] = 2f_1[n] + 4f_1[n - 1]$ and $y_2[n] = 2f_2[n - 2] + 0.5f_2[n - 3]$.
(1) Find the input-output relationship of the system S.
(2) Changing the cascading order of S_1 and S_2, will the input-output relationship of the system S change?

Solution. (1) According to the given conditions, the system can be drawn as in
▶ Figure Q8-5.
Since

$$x[n] = 2f[n] + f[n - 1] \quad \text{and} \quad y[n] = 2x[n - 2] + 0.5x[n - 3],$$

the difference equation of the system can be written in the form

$$y[n] = 4f[n - 2] + 3f[n - 3] + 0.5f[n - 4].$$

(2) It can be proved that if the cascading order of S_1 and S_2 is reversed, the input-output relationship of the system will not change.

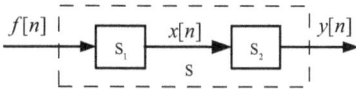

Fig. Q8-5

8.8 Learning tips

Discrete system analysis is another important part after the continuous system in the system analysis. It can be used by referencing all the contents in continuous system analysis. Please pay attention to the following points.
(1) The unit impulse sequence and the step sequence are the basis of time domain analysis.
(2) The convolution sum is a mathematical operation like the convolution integral, but it has various computations and is usually simpler than the convolution integral.
(3) The concept of $h[n]$, similarities and differences between it and $h(t)$ should be borne in mind.

8.9 Problems

Problem 8-1. Test the periodicity of following sequences and find their T if they are periodic.

(1) $f_1[n] = A \cos\left(\frac{n\pi}{4}\right)$

(3) $f_3[n] = A \sin\left(\frac{3n\pi}{4} - \frac{\pi}{4}\right)$

(2) $f_2[n] = A \sin(2n - \pi)$

(4) $f_4[n] = e^{j\left(\frac{n}{8} - \pi\right)}$

Problem 8-2. Please plot the following discrete signals.

(1) $\left(\frac{1}{2}\right)^n \varepsilon[n]$

(5) $\varepsilon[n] + \sin \frac{n\pi}{8} \varepsilon[n]$

(2) $\left(\frac{1}{2}\right)^n \varepsilon[n-2]$

(6) $n \cdot 2^{-n} \varepsilon[n]$

(3) $\left(\frac{1}{2}\right)^{n-2} \varepsilon[n-2]$

(7) $2^n (\varepsilon[-n] - \varepsilon[3-n])$

(4) $2\delta[n] - \varepsilon[n]$

(8) $(n^2 + n + 1)(\delta[n+1] - 2\delta[n])$

Problem 8-3. Knowing $f[n] = n[\varepsilon[n] - \varepsilon[n-7]]$, draw the following signal waveforms.

(1) $f_1[n] = f[-n]$

(4) $f_4[n] = f[n+1] + f[n-1]$

(2) $f_2[n] = f[n+1]$

(5) $f_5[n] = f[n+1]f[n-1]$

(3) $f_3[n] = f[n-1]$

(6) $f_6[n] = f[3n]$

Problem 8-4. $f[n] = \sin \frac{n\pi}{5} [\varepsilon[n] - \varepsilon[n-11]]$ is known. Plot $f[n]$, $f[n-2]$, $f[3-n]$, $\nabla f[n]$ and $\sum_{i=-\infty}^{n} f[i]$.

Problem 8-5. Calculate the first-order backward differences of the following sequences.

(1) $f_1[n] = \varepsilon[n]$

(3) $f_3[n] = n^2 \varepsilon[n]$

(2) $f_2[n] = n\varepsilon[n]$

(4) $f_4[n] = a^n \varepsilon[n]$

Problem 8-6. Write the expressions of the waveforms shown in ▶ Figure P8-6.

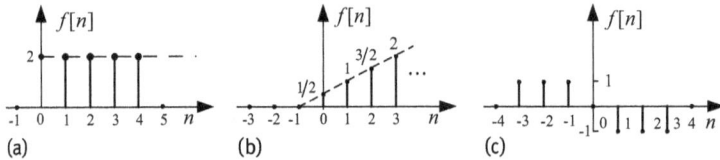

(a)   (b)   (c)   **Fig. P8-6**

Problem 8-7. Find the convolution sum $y[n] = f_1[n] * f_2[n]$ from its definition.

(1) $f_1[n] = \varepsilon[n]$, $f_2[n] = \varepsilon[n]$;

(2) $f_1[n] = 0.5^n \varepsilon[n]$, $f_2[n] = \varepsilon[n]$;

(3) $f_1[n] = \varepsilon[n] - \varepsilon[n-4]$, $f_2[n] = 3\delta[n] + 2\delta[n-1] + \delta[n-2]$;

(4) $f_1[n] = 0.5^n \varepsilon[n]$, $f_2[n] = \varepsilon[n] - \varepsilon[n-5]$.

Problem 8-8. The waveforms of $f_1[n]$ and $f_2[n]$ are shown in ▶ Figure P8-8. Plot $y[n] = f_1[n] * f_2[n]$.

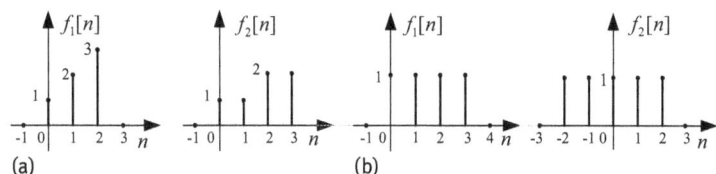

(a) (b) **Fig. P8-8**

Problem 8-9. Prove the following convolution sum properties:

(1) $f[n] * \delta[n] = f[n]$,

(2) $f[n] * \delta[n \pm k] = f[n \pm k]$,

(3) $f[n - k_1] * \delta[n - k_2] = f[n - k_1 - k_2]$,

(4) $\delta[n - k_1] * \delta[n - k_2] = \delta[n - k_1 - k_2]$.

Problem 8-10. $f_1[n] = n[\varepsilon[n] - \varepsilon[n-4]]$ and $f_2[n] = \varepsilon[n+2] - \varepsilon[n-3]$ are given. Find $f_1[n] * f_2[n]$ using the properties of discrete convolution.

Problem 8-11. For a discrete system with nonzero starting state, when the excitation is $f[n]$, the full response is $y_1[n] = [(\frac{1}{2})^n + 1]\varepsilon[n]$; and when the excitation is $-f[n]$, the full response is $y_2[n] = [(-\frac{1}{2})^n - 1]\varepsilon[n]$. Find the full response when the starting state is doubled and the excitation is $4f[n]$.

Problem 8-12. The starting state is zero, if the excitation is $\varepsilon[n]$, the response is $(2^n + 3 \times 5^n + 10)\varepsilon[n]$.

(1) Write the difference equation of the system.

(2) If the excitation is $2(\varepsilon[n] - \varepsilon[n - 10])$, find the zero-state response.

Problem 8-13. From the difference equation write the transfer operator and the unit response for each system.

(1) $y[n] - 2y[n - 1] = f[n]$;

(2) $y[n] - 7y[n - 1] + 10y[n - 2] = f[n] + 2f[n - 1]$,

(3) $y[n] + 3y[n - 1] + 2y[n - 2] = f[n - 1]$.

Problem 8-14. Find the solution for each difference equation using the classic method in the time domain.

(1) $y[n] + \frac{1}{12}y[n - 1] - \frac{1}{12}y[n - 2] = 2^{-n}\varepsilon[n]$, $y[0] = 0$, $y[1] = 0$,

(2) $y[n + 1] + 3y[n] + 2y[n - 1] = (1 + n)(\varepsilon[n] - \varepsilon[n - 1])$, $y[0] = 1$, $y[1] = -3$,

(3) $y[n] + y[n - 2] = f[n]$, $f[n] = \cos\frac{\pi n}{2}\varepsilon[n]$, $y[-1] = y[-2] = 0$.

Problem 8-15. Find the unit responses of the systems described by following difference equations:

(1) $y[n] + \frac{1}{3}y[n-1] = f[n] - 3f[n-1]$,

(2) $y[n+1] + 3y[n] + 2y[n-1] = f[n]$,

(3) $y[n] - \frac{1}{4}y[n-1] = f[n]$,

(4) $y[n] + 0.6y[n-1] - 0.16y[n-2] = f[n]$

(5) $y[n+2] - 0.6y[n+1] - 0.16y[n] = f[n]$,

(6) $y[n+2] - y[n] = f[n+1] - f[n]$.

Problem 8-16. Find the unit response of the system shown in ▶ Figure P8-16.

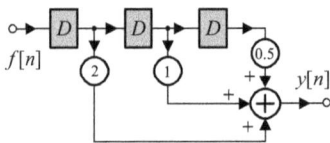

Fig. P8-16

Problem 8-17. Find the unit response and the unit step response of the system shown in ▶ Figure P8-17.

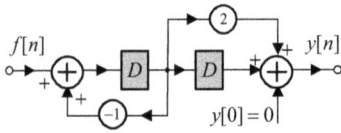

Fig. P8-17

Problem 8-18. Three systems are shown in ▶ Figure P8-18. The unit responses of the subsystems are, respectively, $h_1[n] = \varepsilon[n]$, $h_2[n] = \delta[n-3]$, $h_3[n] = (0.8)^n \varepsilon[n]$. Prove that the three systems are equivalent.

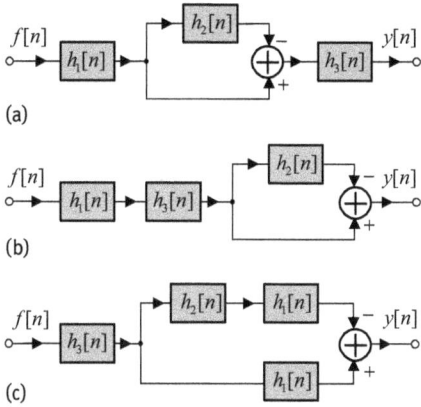

(a)

(b)

(c)   **Fig. P8-18**

Problem 8-19. Find the zero-input responses of the following systems.

(1) $y[n] - 6y[n-1] + 8y[n-2] = f[n]$, $y_x[-1] = \frac{1}{2}$, $y_x[0] = 0$.

(2) $5y[n] - 6y[n-1] = f[n]$, $f[n] = 10\varepsilon[n]$, $y[0] = 1$.

(3) $6y[n] - 5y[n-1] + y[n-2] = f[n]$, $f[n] = (-1)^{n+2}\varepsilon[n-2]$, $y[0] = 15$, $y[1] = 9$.

Problem 8-20. Find the complete response for each system.

(1) $y[n] + y[n-1] - 6y[n-2] = f[n]$, $f[n] = \varepsilon[n]$, $y_x[-2] = \frac{17}{36}$, $y_x[-1] = -\frac{1}{6}$.

(2) $y[n] - 0.1y[n-1] - 0.2y[n-2] = 1.4$, $n \geq 0$, $y_x[0] = 0.6$, $y_x[1] = 0.01$.

(3) $y[n] + 2y[n-1] = f[n]$, $f[n] = e^{-n}\varepsilon[n]$, $y[-1] = 1$.

(4) $y[n] + 2y[n-1] = f[n]$, $f[n] = e^{-n}\varepsilon[n]$, $y[0] = 1$.

Problem 8-21. An LTI system is $y[n] - 0.7y[n-1] + 0.1y[n-2] = 7f[n] - 2f[n-1]$, the input is $f[n] = \varepsilon[n]$ and the initial values are $y[0] = 14$ and $y[1] = 13.1$. Find the full response, free and forced responses, transient and steady state responses.

9 Analysis of discrete signals and systems in the z domain

Questions: After learning analysis methods in *time* domain for discrete signals and systems, readers may ask: how can we analyze the discrete signals and systems in z domain?

How to solve: decompose discrete signal → seek system function → get responses in z domain.

Results: z transform; system function; z transform of zero-state response equals to the product of the system function and the z transform of the excitation sequence.

9.1 Definition of the z transform

Looking back at the analysis methods for continuous systems in the transform domain, we know that their basic thoughts are to decompose the excitation signal into the discrete or continuous sum of some basic signals like $e^{jn\omega_0 t}$, $e^{j\omega t}$ or e^{st} by means of the Fourier series, the Fourier transform and the Laplace transform, then to get the complete response of the system using the linearity and time invariant features. So, can these thoughts be suitable in the analysis of a discrete system? The answer is yes, the decomposition of a discrete signal can be achieved by the z transform, which is the focus in this chapter.

Referring to the defining way of the image function for a continuous signal in the s domain, we can define a transform domain called the z domain for discrete signals firstly. So, for a sequence $f[n]$, its image function or the z transform is defined as

$$F(z) \overset{\text{def}}{=} \sum_{n=-\infty}^{\infty} f[n]z^{-n} , \tag{9.1-1}$$

where $z = re^{j\Omega}$ is a complex number, and the formula is called the bilateral z transform of $f[n]$.

If $f[n]$ is a causal sequence, equation (9.1-1) can be rewritten as

$$F(z) = \sum_{n=0}^{\infty} f[n]z^{-n} . \tag{9.1-2}$$

It is called the unilateral z transform of $f[n]$. Without special instructions, the z transform in this book is the unilateral z transform and is denoted as

$$F(z) = \mathcal{Z}[f[n]] . \tag{9.1-3}$$

From its definition, to ensure the existence of the z transform of $f[n]$, the series on the right side of equation (9.1-1) or equation (9.1-2) should be convergent. This means that

https://doi.org/10.1515/9783110541205-002

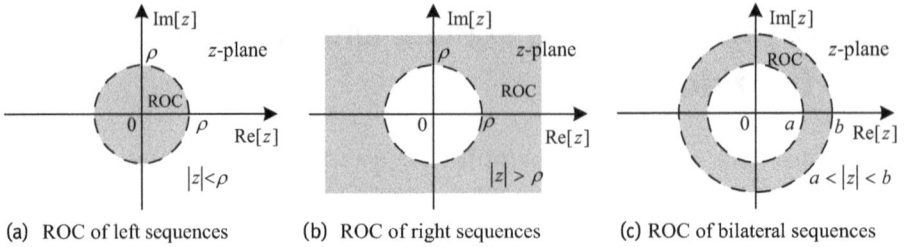

(a) ROC of left sequences   (b) ROC of right sequences   (c) ROC of bilateral sequences

Fig. 9.1: Regions of convergences of sequences.

values of z must be limited in a range which can make the series convergent. According to the series theory, the sufficient condition that makes $F(z)$ convergent is that $f[n]$ is absolutely summable, namely,

$$\sum_{n=-\infty}^{\infty} |f[n]z^{-n}| < \infty \quad \text{(bilateral z transform)}$$

$$\sum_{n=0}^{\infty} |f[n]z^{-n}| < \infty \quad \text{(unilateral z transform)}$$

So, we can define that for any sequence f[n], a set of z values that can make F(z) absolutely summable is called the region of convergence or, simply, ROC.

Here some conclusions can be reached:

(1) The ROC of a reverse causal sequence (left-sided sequence) is inside a circle with the radius ρ, which is called the convergence radius. That means that when $|z| < \rho$ the z transform of the sequence exists.

(2) The ROC of a causal sequence (right-sided sequence) is outside a circle with the radius ρ. This means that when $|z| > \rho$ the z transform of the sequence exists.

(3) The ROC of a bilateral sequence is a ring-shaped area.

(4) The ROC of a finite length sequence is the whole z plane. (except $z = 0$ and $z = \infty$).

(5) The ROC is an open set in any case, that is, it does not include the boundary.

(6) The ROC does not include any poles because the image function is not convergent at poles, and poles are often on the boundary.

Note that the values of ρ relate to $f[n]$. For instance, if $f[n] = a^n$, $n \geq 0$, so $F(z) = \sum_{n=0}^{\infty} a^n z^{-n} = \sum_{n=0}^{\infty} \left(\frac{a}{z}\right)^n$. From series theory, when $|z| > |a|$, the $F(z)$ should be convergent at $\frac{z}{z-a}$, which means $\rho = |a|$.

Based on the above conclusions, the ROCs of those sequences are illustrated in ▶ Figure 9.1. It is necessary to note that for the bilateral z transform, the concept of ROC is very important, and the ROC generally appears together with the expression of the bilateral z transform.

The z transform is given by equation (9.1-2), which is employed to find the image function $F(z)$ via $f[n]$. If an image function $F(z)$ is given, its original function $f[n]$ can be deduced by the following formula:

$$f[n] = \frac{1}{2\pi j} \oint_c F(z)z^{n-1}dz \quad n \geq 0, \tag{9.1-4}$$

where c is a closed circular contour traversed in the counterclockwise direction in the ROC. This expression is called the inverse z transform and is denoted as

$$f[n] = \mathcal{Z}^{-1}[F(z)].$$

Therefore, $f[n]$ and $F(z)$ constitute the z transform pair and can be written as

$$f[n] \overset{\mathcal{Z}}{\longleftrightarrow} F(z). \tag{9.1-5}$$

Equation (9.1-4) can be rewritten as

$$f[n] = \oint_c \left(\frac{1}{2\pi j} \frac{F(z)}{z} \right) z^n dz \quad n \geq 0, \tag{9.1-6}$$

where z^n is a basic signal similar to $e^{j\omega t}$ and e^{st}, while $\frac{1}{2\pi j} \frac{F(z)}{z}$ can be considered as the complex amplitude of z^n.

By comparing equation (9.1-6) with equation (6.1-4), we know that the z transform can decompose the discrete signal $f[n]$ into the continuous sum of the basic sequence z^n, which is similar to the Laplace transform. Thus, this work can achieve our expected purpose and establish the foundation for discrete system analysis.

9.2 z transforms of typical sequences

From the discussion of continuous system analysis, we need to understand some z transforms of basic sequences, so the z transforms of several typical sequences are given in the following examples.

Example 9.2-1. Find the z transform of the sequence shown in ▶ Figure 9.2.

Solution. For this finite length sequence, the image function can be directly solved by the definition of the z transform.

$$F(z) = \mathcal{Z}[f[n]] = \sum_{n=0}^{3} z^{-n} = z^0 + z^{-1} + z^{-2} + z^{-3} = \frac{z^3 + z^2 + z + 1}{z^3}.$$

Its ROC is the whole z plane.

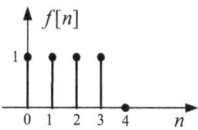

Fig. 9.2: Sequence of E9.2-1.

Example 9.2-2. Find the z transform of the unit sequence $\delta[n]$.

Solution.

$$F(z) = \mathcal{Z}[\delta[n]] = \sum_{n=0}^{\infty} \delta[n]z^{-n} = \delta[0]z^0 = 1,$$

then

$$\delta[n] \overset{z}{\longleftrightarrow} 1.$$

Its ROC is the entire z plane.

Example 9.2-3. Find the z transform for the unit step sequence $\varepsilon[n]$.

Solution.

$$F(z) = \mathcal{Z}[\varepsilon[n]] = \sum_{n=0}^{\infty} \varepsilon[n]z^{-n} = \sum_{n=0}^{\infty} z^{-n},$$

when $|z^{-1}| < 1$ ($|z| > 1$), the series $\sum_{n=0}^{\infty} z^{-n}$ is convergent, so

$$F(z) = \frac{1}{1 - z^{-1}} = \frac{z}{z - 1},$$

and then, we have

$$\varepsilon[n] \overset{z}{\longleftrightarrow} \frac{z}{z - 1}.$$

Example 9.2-4. Find the z transform of the exponential sequence $a^n \varepsilon[n]$.

Solution.

$$F(z) = \mathcal{Z}[a^n \varepsilon[n]] = \sum_{n=0}^{\infty} a^n z^{-n} = \sum_{n=0}^{\infty} \left(\frac{a}{z}\right)^n.$$

Tab. 9.1: z transforms of some common unilateral sequences.

No.	$f[n]$	$F(z)$	ROC
1	$\delta[n]$	1	$|z| \geq 0$
2	$\varepsilon[n]$	$\frac{z}{z-1}$	$|z| > 1$
3	$n\varepsilon[n]$	$\frac{z}{(z-1)^2}$	$|z| > 1$
4	$n^2 \varepsilon[n]$	$\frac{z(z+1)}{(z-1)^2}$	$|z| > 1$
5	$a^n \varepsilon[n]$	$\frac{z}{z-a}$	$|z| > |a|$
6	$na^{n-1}\varepsilon[n]$	$\frac{z}{(z-a)^2}$	$|z| > |a|$
7	$e^{an}\varepsilon[n]$	$\frac{z}{z-e^a}$	$|z| > |e^a|$
8	$\cos[\beta n]\varepsilon[n]$	$\frac{z(z-\cos\beta)}{z^2 - 2z\cos\beta + 1}$	$|z| > 1$
9	$\sin[\beta n]\varepsilon[n]$	$\frac{z\sin\beta}{z^2 - 2z\cos\beta + 1}$	$|z| > 1$
10	$e^{-an}\cos[\beta n]\varepsilon[n]$	$\frac{z(z-e^{-a}\cos\beta)}{z^2 - 2ze^{-a}\cos\beta + e^{-2a}}$	$|z| > |e^{-a}|$
11	$e^{-an}\sin[\beta n]\varepsilon[n]$	$\frac{ze^{-a}\sin\beta}{z^2 - 2ze^{-a}\cos\beta + e^{-2a}}$	$|z| > |e^{-a}|$

When $\left|\frac{a}{z}\right| < 1$ ($|z| > |a|$), the series $\sum_{n=0}^{\infty} \left(\frac{a}{z}\right)^n$ is convergent, so

$$F(z) = \frac{1}{1 - az^{-1}} = \frac{z}{z - a},$$

and then

$$a^n \varepsilon[n] \xleftrightarrow{z} \frac{z}{z - a}.$$

The z transforms for some commonly used unilateral sequences are listed in Table 9.1.

9.3 Properties of the z transform

1. Linearity

If

$$f_1[n] \xleftrightarrow{z} F_1(z) \quad \text{and} \quad f_2[n] \xleftrightarrow{z} F_2(z),$$

then

$$af_1[n] + bf_2[n] \xleftrightarrow{z} aF_1(z) + bF_2(z), \tag{9.3-1}$$

where both a and b are arbitrary constants.

2. Time shifting

If

$$f[n] \xleftrightarrow{z} F(z) \quad \text{and} \quad k > 0,$$

then

(1)

$$f[n - k]\varepsilon[n] \xleftrightarrow{z} z^{-k}\left[F(z) + \sum_{m=-k}^{-1} f[m]z^{-m}\right], \tag{9.3-2}$$

(2)

$$f[n - k]\varepsilon[n - k] \xleftrightarrow{z} z^{-k}F(z), \tag{9.3-3}$$

(3)

$$f[n + k]\varepsilon[n] \xleftrightarrow{z} z^{k}\left[F(z) - \sum_{m=0}^{k-1} f[m]z^{-m}\right]. \tag{9.3-4}$$

Example 9.3-1. Prove the time shifting properties based on the sequence inbreak ▶ Figure 9.3a.

Proof. (1) According to the definition of the z transform,

$$\mathcal{Z}[f[n - k]\varepsilon[n]] = \sum_{n=0}^{\infty} f[n - k]z^{-n} = \sum_{n=0}^{\infty} f[n - k]z^{-(n-k)}z^{-k}$$

$$\overset{m=n-k}{=} z^{-k}\sum_{m=-k}^{\infty} f[m]z^{-m} = z^{-k}\sum_{m=0}^{\infty} f[m]z^{-m} + z^{-k}\sum_{m=-k}^{-1} f[m]z^{-m}$$

$$= z^{-k}F(z) + z^{-k}\sum_{m=-k}^{-1} f[m]z^{-m} = z^{-k}\left[F(z) + \sum_{m=-k}^{-1} f[m]z^{-m}\right]$$

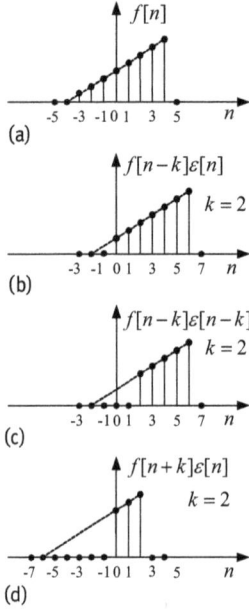

(a)

(b)

(c)

(d)

Fig. 9.3: A sequence and its time shifting sequences.

(2) According to the definition of the z transform,

$$\mathcal{Z}[f[n-k]\varepsilon[n-k]] = \sum_{n=0}^{\infty} f[n-k]\varepsilon[n-k]z^{-n} = \sum_{n=0}^{\infty} f[n-k]\varepsilon[n-k]z^{-(n-k)}z^{-k}$$

$$\overset{m=n-k}{=} z^{-k}\sum_{m=-k}^{\infty} f[m]\varepsilon[m]z^{-m} = z^{-k}\sum_{m=0}^{\infty} f[m]z^{-m} = z^{-k}F(z)$$

(3) According to the definition of the z transform,

$$\mathcal{Z}[f[n+k]\varepsilon[n]] = \sum_{n=0}^{\infty} f[n+k]z^{-n} = \sum_{n=0}^{\infty} f[n+k]z^{-(n+k)}z^{k}$$

$$\overset{m=n+k}{=} z^{k}\sum_{m=k}^{\infty} f[m]z^{-m} = z^{k}\sum_{m=0}^{\infty} f[m]z^{-m} - z^{k}\sum_{m=0}^{k-1} f[m]z^{-m}$$

$$= z^{k}F(z) - z^{k}\sum_{m=0}^{k-1} f[m]z^{-m} = z^{-k}\left[F(z) - \sum_{m=0}^{k-1} f[m]z^{-m}\right]$$

□

3. Scaling in the z domain

If

$$f[n] \overset{z}{\longleftrightarrow} F(z),$$

then

$$a^{n}f[n] \overset{z}{\longleftrightarrow} F\left(\frac{z}{a}\right), \tag{9.3-5}$$

Proof.

$$Z[a^n f[n]] = \sum_{n=0}^{\infty} a^n f[n] z^{-n} = \sum_{n=0}^{\infty} f[n] \left(\frac{z}{a}\right)^{-n} = F\left(\frac{z}{a}\right).$$

□

4. Convolution sum property

If

$$f_1[n]\varepsilon[n] \xrightarrow{Z} F_1(z) \quad \text{and} \quad f_2[n]\varepsilon[n] \xrightarrow{Z} F_2(z),$$

then

$$(f_1[n]\varepsilon[n]) * (f_2[n]\varepsilon[n]) \xrightarrow{Z} F_1(z)F_2(z). \tag{9.3-6}$$

Proof. According to the definition of convolution sum,

$$(f_1[n]\varepsilon[n]) * (f_2[n]\varepsilon[n]) = \sum_{k=-\infty}^{\infty} f_1[k]\varepsilon[k]f_2[n-k]\varepsilon[n-k] = \sum_{k=0}^{\infty} f_1[k]f_2[n-k]\varepsilon[n-k]$$

According to the definition of the z transform,

$$Z[f_1[n]\varepsilon[n] * f_2[n]\varepsilon[n]] = \sum_{n=0}^{\infty} \left(\sum_{k=0}^{\infty} f_1[k]f_2[n-k]\varepsilon[n-k]\right)z^{-n}$$

$$\underset{=}{\overset{\text{change}}{\underset{\text{summing order}}{}}} \sum_{k=0}^{\infty} f_1[k] \sum_{n=0}^{\infty} f_2[n-k]\varepsilon[n-k]z^{-n}$$

$$\underset{=}{\overset{\text{time}}{\underset{\text{shifting}}{}}} \sum_{k=0}^{\infty} f_1[k]z^{-k}F_2(z) = F_1(z)F_2(z)$$

□

In addition to the above properties, the main properties of the z transform are listed in Table 9.2.

Note:

(1) Values of a sequence are considered as zero at points of which n/k is not an integer, for property 4 (scaling in the time domain).

(2) All the poles of $F(z)$ must be inside the unit circle. If there is a pole on the unit circle, it is only located at the point $z = +1$ and is of first order, for the property 11 (terminal value theorem).

Example 9.3-2. Find the z transform of a unilateral periodic unit sequence with period N.

$$\delta_N[n]\varepsilon[n] = \delta[n] + \delta[n-N] + \delta[n-2N] + \cdots + \delta[n-mN] + \cdots = \sum_{m=0}^{\infty} \delta[n-mN]$$

Tab. 9.2: Properties of the unilateral z transform.

No.	Name	Time domain	z domain
1	Linearity	$af_1[n] + bf_2[n]$	$aF_1(z) + bF_2(z)$
2	Time shifting $k > 0$	$f[n - k]\varepsilon[n]$	$z^{-k}[F(z) + \sum_{m=-k}^{-1} f(m)z^{-m}]$
		$f[n + k]\varepsilon[n]$	$z^k[F(z) - \sum_{m=0}^{k-1} f(m)z^{-m}]$
		$f[n - k]\varepsilon[n - k]$	$z^{-k}F(z)$
3	Scaling in the z domain	$a^n f[n]$	$F(\frac{z}{a})$
4	Scaling in the time domain	$f[\frac{n}{k}], \quad k = 1, 2, 3, \ldots$	$F(z^k)$
5	Convolution sum in the time domain	$f_1[n]\varepsilon[n] * f_2[n]\varepsilon[n]$	$F_1(z)F_2(z)$
6	Multiplication in the time domain	$f_1[n] \cdot f_2[n]$	$\frac{1}{2\pi j} \oint_c \frac{F_1(\eta)F_2(\frac{z}{\eta})}{\eta} d\eta$
7	Partial sum in the time domain	$\sum_{k=0}^{n} f[k]$	$\frac{z}{z-1} F(z)$
8	Differentiation in the z domain	$n^k f[n]$	$-z \frac{d^k F(z)}{dz^k}$
9	Integration in the z domain	$\frac{f[n]}{n+k} \quad n + k > 0$	$z^k \int_z^\infty \frac{F(\eta)}{\eta^{k+1}} d\eta$
10	Initial value theorem	$f[0] = \lim_{z \to \infty} F(z)$	
11	Terminal value theorem	$f[\infty] = \lim_{n \to \infty} f[n] = \lim_{z \to 1}(z - 1)F(z)$	

Solution. Because

$$\delta[n] \overset{z}{\longleftrightarrow} 1 \,,$$

from the time shifting property, the z transforms of right-shifted sequences are

$$\delta[n - N]\varepsilon[n - N] \overset{z}{\longleftrightarrow} z^{-N} \,,$$

$$\delta[n - 2N]\varepsilon[n - 2N] \overset{z}{\longleftrightarrow} z^{-2N} \,,$$

$$\vdots$$

$$\delta[n - mN]\varepsilon[n - mN] \overset{z}{\longleftrightarrow} z^{-mN} \,.$$

$$\vdots$$

Then, the z transform of a unilateral periodic unit sequence is

$$F(z) = \mathcal{Z}[\delta_N[n]\varepsilon[n]] = 1 + z^{-N} + z^{-2N} + \cdots + z^{-mN} + \cdots$$

$$= \frac{1}{1 - z^{-N}} = \frac{z^N}{z^N - 1}, \quad |z^N| > 1 \,.$$

Example 9.3-3. Find the z transforms of the following sequences by the properties of the z transform.

(1) $\sin[\beta n]\varepsilon[n]$ (2) $(-1)^n n\varepsilon[n]$ (3) $\sum_{i=0}^{n} \left(-\frac{1}{2}\right)^i$ (4) $\frac{a^n}{n+1}\varepsilon[n]$

Solution. (1) According to Euler's relation, we have

$$\sin[\beta n]\varepsilon[n] = \frac{1}{2j}\left[e^{j\beta}\varepsilon[n] - e^{-j\beta}\varepsilon[n]\right].$$

With the scaling transform in the z domain and linearity, we have

$$\mathcal{Z}[\sin[\beta n]\varepsilon[n]] = \frac{1}{2j}\left[\frac{\frac{z}{e^{j\beta}}}{\frac{z}{e^{j\beta}}-1} - \frac{\frac{z}{e^{-j\beta}}}{\frac{z}{e^{-j\beta}}-1}\right] = \frac{z}{2j}\left[\frac{1}{z-e^{j\beta}} - \frac{1}{z-e^{-j\beta}}\right]$$

$$= \frac{z}{2j}\left[\frac{e^{j\beta} - e^{-j\beta}}{z^2 - z\left(e^{j\beta} + e^{-j\beta}\right) + 1}\right] = \frac{z\sin\beta}{z^2 - 2z\cos\beta + 1}.$$

(2) If $f_1[n] = n\varepsilon[n]$, according to the differential property in the z domain, we have

$$F_1(z) = \mathcal{Z}[n\varepsilon[n]] = -z\frac{d}{dz}\left(\frac{z}{z-1}\right) = \frac{z}{(z-1)^2}.$$

From the scaling transform in the z domain, we have

$$\mathcal{Z}[(-1)^n n\varepsilon[n]] = F_1(-z) = \frac{-z}{(z+1)^2}, \quad |z| > 1.$$

(3) Because

$$\left(-\frac{1}{2}\right)^n \varepsilon[n] \overset{\mathcal{Z}}{\longleftrightarrow} \frac{z}{z+\frac{1}{2}},$$

with the partial sum property in the time domain, we have

$$\mathcal{Z}\left[\sum_{i=0}^{n}\left(-\frac{1}{2}\right)^i\right] = \frac{z}{z-1} \cdot \frac{2z}{2z+1} = \frac{z^2}{(z-1)\left(z+\frac{1}{2}\right)}, \quad |z| > \frac{1}{2}.$$

(4) If $f_1[n] = a^n\varepsilon[n]$, we have

$$F_1(z) = \mathcal{Z}[a^n\varepsilon[n]] = \frac{z}{z-a}.$$

With integration property in the z domain, we have

$$\mathcal{Z}\left[\frac{a^n}{n+1}\varepsilon[n]\right] = z\int_{z}^{\infty}\frac{F_1(x)}{x^2}dx = z\int_{z}^{\infty}\frac{1}{x(x-a)}dx = \frac{z}{a}\ln\frac{z}{z-a}, \quad |z| > a.$$

9.4 Solutions of the inverse z transform

The procedure to obtain the original function $f[n]$ from its image function $F(z)$ is called the inverse z transform. Usually, there are three ways to achieve this aim:
(1) the power series expansion method;
(2) the partial fraction expansion method;
(3) the residue method (contour integral method).

We will introduce mainly the first two methods in this book.

1. Power series expansion method

We can expand the definition expression of the z transform as

$$F(z) = \sum_{n=0}^{\infty} f[n]z^{-n} = f[0]z^0 + f[1]z^{-1} + f[2]z^{-2} + f[3]z^{-3} + \cdots \qquad (9.4\text{-}1)$$

It can be seen that $F(z)$ can be expanded into a negative power series of z by the z transform definition. Moreover, the coefficient of each term z^{-n} is a function value of the corresponding original function $f[n]$.

Example 9.4-1. Use the power series expansion method to verify $\frac{1}{n}a^n \overset{z}{\longleftrightarrow} \ln \frac{z}{z-a}$.

Proof. According to the series theories, when $|x| < 1$,

$$\ln(1 - x) = -\left(x + \frac{x^2}{2} + \frac{x^3}{3} + \cdots + \frac{x^n}{n} + \cdots \right).$$

If $x = \frac{a}{z}$, we have $\ln(1 - x) = -\ln \frac{z}{z-a}$. So,

$$\ln \frac{z}{z - a} = \frac{a}{z} + \frac{1}{2}\left(\frac{a}{z}\right)^2 + \frac{1}{3}\left(\frac{a}{z}\right)^3 + \cdots + \frac{1}{n}\left(\frac{a}{z}\right)^n + \cdots$$

$$= az^{-1} + \frac{1}{2}a^2z^{-2} + \frac{1}{3}a^3z^{-3} + \cdots + \frac{1}{n}a^nz^{-n} + \cdots$$

Therefore, the original function is

$$f[n] = \frac{1}{n}a^n,$$

namely,

$$\frac{1}{n}a^n \overset{z}{\longleftrightarrow} \ln \frac{z}{z - a}, \quad |z| > a.$$

From this example, we can say that if $F(z)$ is given in fractional form, it can be expanded into a power series by long division. Then its inverse z transform can be obtained. This method is applicable when $f[n]$ cannot be written as a simple analytic expression or when only a few values of $f[n]$ are required to be found. □

Example 9.4-2. Known is $F(z) = \frac{z^3+2z^2+1}{z^3-1.5z^2+0.5z}$, $|z| > 1$. Find its original sequence $f[n]$.

Solution. $F(z)$ can be expanded into a power series using long division,

$$F(z) = 1 + 3.5z^{-1} + 4.75z^{-2} + 6.375z^{-3} + \cdots,$$

so, we can write

$$f[n] = [\underset{\uparrow}{1}, \ 3.5, \ 4.75, \ 6.375, \ \ldots].$$

2. Partial fraction expansion method

Similarly to the solution of the inverse Laplace transform, we can also obtain the inverse z transform by using the partial fraction expansion method. The core of this method is to decompose $F(z)$ into the algebraic sum of several basic forms first, and then the inverse transform of $F(z)$ can be obtained by the known basic forms. However, it should be noted that the partial fraction expansion of $F(z)$ is different from that of $F(s)$. We can directly carry out the partial fraction expansion of $F(s)$, but for $F(z)$ we must first carry out the partial fraction expansion of $\frac{F(z)}{z}$ and then remove the z in the denominator on the right side of the equation, so that it can ensure that the numerator contains z in each fraction. The reason is that most numerators of fractions include the variable z in the z transforms of the basic sequences in Table 9.1. If $F(z)$ is expanded directly, the term z will not appear in the numerators of fractions, so we cannot get the inverse transforms by these known z transforms.

Example 9.4-3. Find the inverse z transform of the image function $F(z) = \frac{z+2}{2z^2-7z+3}$.

Solution.

$$\frac{F(z)}{z} = \frac{z+2}{z(2z-1)(z-3)} = \frac{k_1}{z} + \frac{k_2}{2z-1} + \frac{k_3}{z-3},$$

$$k_1 = z\left[\frac{F(z)}{z}\right]\Big|_{z=0} = \frac{z+2}{(2z-1)(z-3)}\Big|_{z=0} = \frac{2}{3},$$

$$k_2 = (2z-1)\left[\frac{F(z)}{z}\right]\Big|_{z=\frac{1}{2}} = \frac{z+2}{z(z-3)}\Big|_{z=\frac{1}{2}} = -2,$$

$$k_3 = (z-3)\left[\frac{F(z)}{z}\right]\Big|_{z=3} = \frac{z+2}{z(2z-1)}\Big|_{z=3} = \frac{1}{3}.$$

Therefore,

$$\frac{F(z)}{z} = \frac{2}{3}\cdot\frac{1}{z} - 2\cdot\frac{1}{2z-1} + \frac{1}{3}\cdot\frac{1}{z-3},$$

$$F(z) = \frac{2}{3} - 2\cdot\frac{z}{2z-1} + \frac{1}{3}\cdot\frac{z}{z-3}.$$

So, from Table 9.1 the original sequence is

$$f[n] = \frac{2}{3}\delta[n] - \left(\frac{1}{2}\right)^n + \frac{1}{3}3^n \quad n \geq 0.$$

Example 9.4-4. Find the original sequence of the image function $Y(z) = \frac{z^3+2z^2+1}{z(z-0.5)(z-1)}$.

Solution.

$$\frac{Y(z)}{z} = \frac{z^3 + 2z^2 + 1}{z^2(z - 0.5)(z - 1)} = \frac{k_{01}}{z^2} + \frac{k_{02}}{z} + \frac{k_1}{z - 0.5} + \frac{k_2}{z - 1},$$

$$k_{01} = z^2 \left[\frac{y(z)}{z} \right]\Big|_{z=0} = \frac{z^3 + 2z^2 + 1}{(z - 0.5)(z - 1)}\Big|_{z=0} = 2,$$

$$k_{02} = \frac{d}{dz} \left[z^2 \frac{y(z)}{z} \right]\Big|_{z=0},$$

$$= \frac{(3z^2 + 4z)(z - 0.5)(z - 1) - (z^3 + 2z^2 + 1)(2z - 1.5)}{(z - 0.5)^2(z - 1)^2}\Big|_{z=0} = 6$$

$$k_1 = (z - 0.5)\left[\frac{y(z)}{z} \right]\Big|_{z=0.5} = \frac{z^3 + 2z^2 + 1}{z^2(z - 1)}\Big|_{z=0.5} = -13,$$

$$k_2 = (z - 1)\left[\frac{y(z)}{z} \right]\Big|_{z=1} = \frac{z^3 + 2z^2 + 1}{z^2(z - 0.5)}\Big|_{z=1} = 8.$$

Therefore

$$\frac{Y(z)}{z} = \frac{2}{z^2} + \frac{6}{z} - \frac{13}{z - 0.5} + \frac{8}{z - 1},$$

and then

$$Y(z) = \frac{2}{z} + 6 - \frac{13z}{z - 0.5} + \frac{8z}{z - 1}.$$

From Table 9.1, the original sequence is

$$y[n] = 2\delta[n - 1] + 6\delta[n] - 13(0.5)^n\varepsilon[n] + 8\varepsilon[n] \quad n \geq 0.$$

Note: From the time shifting property, the inverse z transform of $\frac{2}{z}$ is $2\delta[n - 1]$.

9.5 Relations between the z domain and the s domain

We know that $s = \sigma + j\omega$ in the s domain, and $z = re^{j\Omega}$ and $\Omega = \omega T$ in the z domain, where T is the sampling interval for a discrete signal. Both s and z are complex numbers, but their expressed forms are different, because one is the algebraic form and the other is the exponential. Obviously, there should be mapping relationships between them. For example, let their real parts be zero, then there are $\sigma = 0$ and $s = j\omega$ in the s domain, $r = 1$ and $z = e^{j\Omega}$ in the z domain. The mapping relationships between them are shown in Table 9.3.

The unit circle in the z plane can be considered a circle which is bent by the imaginary axis in the s plane in the counterclockwise direction. The left half-plane of the s domain corresponds to the inner region of the unit circle in the z plane, while the right half-plane of the s domain corresponds to the outside region of the unit circle in the z plane. Therefore, since the Fourier transform of a continuous signal is equal to its corresponding Laplace transform on the imaginary axis $j\omega$, we can say that the Fourier transform of a discrete signal is equal to its z transform on the unit circle.

Tab. 9.3: Mapping relationships between the s plane and the z plane.

s plane ($s = \sigma + j\omega$)	z plane ($z = re^{j\Omega}$)
imaginary axis $j\omega$ ($\sigma=0, s=j\omega$)	Im[z] unit circle ($r = 1$, arbitrary Ω)
left-open-half plane $j\omega$ ($\sigma<0$)	Im[z] Internal of unit circle ($r < 1$, arbitrary Ω)
right-open-half space $j\omega$ ($\sigma>0$)	Im[z] outside of unit circle ($r > 1$, arbitrary Ω)
straight line paralleled to the imaginary axis (σ = constant) ($\sigma_1 < 0, \sigma_2 > 0$)	Im[z] circle ($r_1 < 1$) ($r_2 > 1$)
real axis ($\omega=0$, $s=\sigma$)	Im[z] positive real axis ($\Omega = 0$, arbitrary r)

To study clearly and deeply the characteristics of the two planes, the relationships between the ROCs of the bilateral Laplace transform and the z transform are listed in Table 9.4.

Tab. 9.4: The comparison of two kinds of region of convergence.

No.	Bilateral Laplace transform ROC	Bilateral z transform ROC	Signal or sequence in the time domain
1	Limited s plane	Limited z plane	Finite length signals or sequences
2	A half limited s plane on the right of a straight line parallel to the imaginary axis	Outside a circle with the origin as the center in the z plane	Right-side signals or sequences
3	A half limited s plane on the left of a straight line parallel to the imaginary axis.	Inside a circle with the origin as the center in z plane	Left-side signals or sequences
4	Zonal region with limited left and right boundary in the s plane	A ring with an inner and outer finite radius centered at the origin in the z plane	Bilateral signals or sequences

In summary, the z domain is also the complex frequency domain and has no physical meaning, and it is just a mathematical method like the s domain.

The following examples can help us to better understand the similarities and differences between the z and s domains.

Example 9.5-1. Find the unilateral z Transforms of the following sequences, and give the pole-zero diagrams and the regions of convergence.
(1) $f[n] = \{1, -1, 1, -1, 1, \ldots\}$;
(2) $f[n] = \left(\frac{1}{2}\right)^{|n|}$;
(3) $f[n] = \left(\frac{1}{2}\right)^{n} \varepsilon[-n]$

Solution. (1) $f[n] = (-1)^{n}\varepsilon[n]$, and according to the definition of the z transform, we have

$$F(z) = \sum_{n=0}^{\infty}(-1)^{n}\varepsilon(n)z^{-n} = \sum_{n=0}^{\infty}\left(-\frac{1}{z}\right)^{n} = \frac{1}{1 - \left(-\frac{1}{z}\right)} = \frac{z}{1 + z}.$$

The pole is $\lambda = -1$, the zero is $z = 0$, and ROC is $|z| > 1$.

(2) The original sequence can be written as the sum of a left-sided and a right-sided sequence,

$$f[n] = \left(\frac{1}{2}\right)^{n}\varepsilon[n] + \left(\frac{1}{2}\right)^{-n}\varepsilon[-n - 1],$$

$$F(z) = \sum_{n=0}^{\infty}\left[\left(\frac{1}{2}\right)^{n}\varepsilon(n) + \left(\frac{1}{2}\right)^{-n}\varepsilon(-n - 1)\right]z^{-n} = \sum_{n=0}^{\infty}\left(\frac{1}{2}\right)^{n}z^{-n} = \frac{z}{z - \frac{1}{2}}.$$

The pole is $\lambda = \frac{1}{2}$, the zero is $z = 0$, and ROC is $|z| > \frac{1}{2}$.

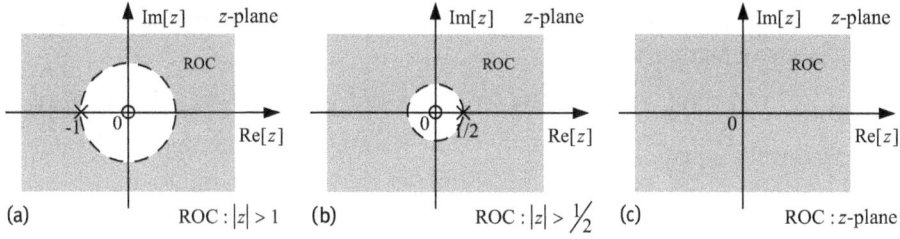

Fig. 9.4: E9.5-1.

(3) The original sequence is a left-sided sequence, and when $n > 0$, its values are all zero, so,

$$F(z) = \sum_{n=0}^{\infty} \left[\left(\frac{1}{2} \right)^n \varepsilon(-n) \right] z^{-n} = 1 \, .$$

There are no poles and no zeros, and ROC is the whole z plane.

The pole-zero diagrams and the ROCs of the above three functions are shown in ▶ Figure 9.4.

9.6 Analysis of discrete systems in the z domain

The reason for introducing the z transform in the analysis to discrete signals is to simplify the analysis methods and the processes, and to provide a shortcut for the analysis to discrete systems, just like the introduction of the Laplace transform in the analysis of continuous systems.

The analysis methods for discrete systems in the z domain are similar to those for continuous systems in the s domain. We can transform a difference equation into an algebraic equation via the z transform for analyzing a discrete system. Of course, we can also employ the system function for it, such as discussing the zero-state response, stability, frequency response, etc.

9.6.1 Analysis method from the difference equation

We can learn the solving process of a difference equation in the z domain by the following example.

Example 9.6-1. The difference equation of an LTI system is

$$y[n] - 3y[n-1] + 2y[n-2] = f[n] + f[n-1]$$

and the starting conditions are $y_x[-2] = 3$, $y_x[-1] = 2$ and $f[n] = 2^n \varepsilon[n]$. Find the zero-input response and the zero-state response of the system.

Solution. If $\mathcal{Z}[y[n]] = Y(z)$ and $\mathcal{Z}[f[n]] = F(z)$, with the z transform on both sides of the difference equation, we have

$$Y(z) - 3z^{-1}Y(z) - 3y[-1] + 2z^{-2}Y(z) + 2y[-2] + 2z^{-1}y[-1] = F(z) + z^{-1}F(z) + f[-1].$$

Rearranging the equation,

$$\left(1 - 3z^{-1} + 2z^{-2}\right)Y(z) = \left(3 - 2z^{-1}\right)y[-1] - 2y[-2] + F(z) + z^{-1}F(z) + f[-1].$$

Obviously, the difference equation is transformed into an algebraic equation by the z transform, so we have

$$Y(z) = \frac{(3 - 2z^{-1})y(-1) - 2y(-2)}{(1 - 3z^{-1} + 2z^{-2})} + \frac{F(z) + z^{-1}F(z) + f(-1)}{(1 - 3z^{-1} + 2z^{-2})},$$

where the first term only relates to the starting state of the system, and it is the image function $Y_x(z)$ of the zero-input response. The second only relates to the system excitation, and it is the image function $Y_f(z)$ of the zero-input response. The excitation is $f[n] = 2^n\varepsilon[n]$, so $y[-2] = y_x[-2] = 3$, $y[-1] = y_x[-1] = 2$ and $f[-1] = 0$, $F(z) = \frac{z}{z-2}$. Then the image functions of the zero-input response and zero-state response are, respectively,

$$Y_x(z) = \frac{(3 - 2z^{-1})y(-1) - 2y(-2)}{1 - 3z^{-1} + 2z^{-2}} = \frac{2(3 - 2z^{-1}) - 6}{1 - 3z^{-1} + 2z^{-2}} = \frac{-4z}{z^2 - 3z + 2},$$

$$Y_f(z) = \frac{F(z) + z^{-1}F(z) + f(-1)}{1 - 3z^{-1} + 2z^{-2}} = \frac{z^2 + z}{z^2 - 3z + 2} \cdot \frac{z}{z - 2}.$$

Expanding $\frac{Y_x(z)}{z}$ and $\frac{Y_f(z)}{z}$ into partial fractions, we obtain

$$\frac{Y_x(z)}{z} = \frac{4}{z - 1} - \frac{4}{z - 2},$$

$$\frac{Y_f(z)}{z} = \frac{6}{(z - 2)^2} - \frac{1}{z - 2} + \frac{2}{z - 1},$$

and

$$Y_x(z) = \frac{4z}{z - 1} - \frac{4z}{z - 2},$$

$$Y_f(z) = \frac{6z}{(z - 2)^2} - \frac{z}{z - 2} + \frac{2z}{z - 1}.$$

Therefore, the zero-input response and zero-state response of the system are

$$y_x[n] = 4(1 - 2^n) \quad n \geq 0,$$

$$y_f[n] = (3n - 1)2^n + 2 \quad n \geq 0.$$

The full response is

$$y[n] = y_x[n] + y_f[n] = 4(1 - 2^n) + (3n - 1)2^n + 2 = (3n - 5)2^n + 6 \quad n \geq 0.$$

So, this result is consistent with that obtained with the method in the time domain, which can be seen in Example 8.5-9.

In general, the difference equation for an LTI causal system can be expressed in the form

$$\sum_{k=0}^{N} a_{N-k}y[n-k] = \sum_{r=0}^{M} b_{M-r}f[n-r] . \tag{9.6-1}$$

With the z transform on both sides of the equation, and from the time shifting property, we have

$$\sum_{k=0}^{N} a_{N-k}z^{-k}\left[Y(z) + \sum_{l=-k}^{-1} y[l]z^{-l} \right] = \sum_{r=0}^{M} b_{M-r}z^{-r}\left[F(z) + \sum_{m=-r}^{-1} f[m]z^{-m} \right] . \tag{9.6-2}$$

For the causal input sequence, we have

$$\sum_{m=-r}^{-1} f[m]z^{-m} = 0 .$$

Hence, we obtain

$$\sum_{k=0}^{N} a_{N-k}z^{-k}\left[Y(z) + \sum_{l=-k}^{-1} y[l]z^{-l} \right] = \sum_{r=0}^{M} b_{M-r}z^{-r}F(z) . \tag{9.6-3}$$

The equation is the z domain relation between the complete response $y[n]$ and the excitation $f[n]$ of a system under nonzero starting conditions. Note that this conclusion is similar to solving the system equation with the Laplace transform, which can give the complete solution of the equation in one fell swoop.

If values of the input sequence are zero, the system response only contains zero-input response $y_x[n]$, whose z transform $Y_x(z)$ satisfies

$$\sum_{k=0}^{N} a_{N-k}z^{-k}\left[Y_x(z) + \sum_{l=-k}^{-1} y_x[l]z^{-l} \right] = 0 . \tag{9.6-4}$$

Obviously, $Y_x(z)$ only depends on the starting state values of zero-input response $y_x[l]$, $l = -1, -2, \ldots, -k$.

Note: Generally speaking, the starting state values of system are denoted as $y[l]$, $l = -1, -2, \ldots, -k$, and are not the same as $y_x[l]$. However, for a causal input sequence, because $f[n] = 0$ for $n < 0$, so $y[l] = y_x[l], l = -1, -2, \ldots, -k$.

If the starting state values are $y[l] = 0$, $l = -1, -2, \ldots, -k$, the response $y[n]$ is equal to the zero-state response $y_f[n]$, so $Y_f(z)$ meets

$$\sum_{k=0}^{N} a_{N-k}z^{-k}Y_f(z) = \sum_{r=0}^{M} b_{M-r}z^{-r}F(z) . \tag{9.6-5}$$

Obviously, $Y_f(z)$ only relate to the input sequence $f[n]$ of the system.

Observing equations (9.6-3)–(9.6-5) carefully, we find

$$Y(z) = Y_x(z) + Y_f(z) \tag{9.6-6}$$

That is, the z transform of the complete response also meets the decomposition. This conclusion is the same as the one in the Laplace transform.

In summary, we have given two analysis methods for discrete systems by solving the difference equation in the z domain:

(1) Obtain the z transform $Y(z)$ of the complete response $y[n]$ by equation (9.6-3) first, then obtain the expression of $y[n]$ by finding inverse z transform of $Y(z)$.

(2) First, find $Y_x(z)$ and $Y_f(z)$ by equations (9.6-4) and (9.6-5), then obtain the corresponding $y_x[n]$ and $y_f[n]$ with the inverse z transform, and finally, add them to form the complete response $y[n]$.

The following example will show how obtain the complete response of a system expressed by a forward difference equation in the z domain.

Example 9.6-2. $y[0] = 1$ and $y[1] = 2$ are known. Find the full response of the following system by the analysis method in the z domain,

$$y[n + 2] + y[n + 1] + y[n] = \varepsilon[n]$$

Solution. With the z transform on both sides of the difference equation, we obtain

$$z^2 Y(z) - z^2 y[0] - zy[1] + zY(z) - zy[0] + Y(z) = \frac{z}{z - 1}.$$

Putting the initial conditions $y[0] = 1$ and $y[1] = 2$ into the equation, yields

$$Y(z) = \frac{z}{(z - 1)(z^2 + z + 1)} + \frac{z^2 + 3z}{z^2 + z + 1}. \tag{9.6-7}$$

We note that the denominator contains $z^2 + z + 1$, so using the partial fraction expansion method find the inverse z transform may be difficult. In Table 9.1, the z transforms of $\cos[\beta n]\varepsilon[n]$ and $\sin[\beta n]\varepsilon[n]$ are, respectively,

$$\cos[\beta n]\varepsilon[n] \overset{z}{\longleftrightarrow} \frac{z(z - \cos \beta)}{z^2 - 2z \cos \beta + 1}, \tag{9.6-8}$$

$$\sin[\beta n]\varepsilon[n] \overset{z}{\longleftrightarrow} \frac{z \sin \beta}{z^2 - 2z \cos \beta + 1}. \tag{9.6-9}$$

If we let $-2z \cos \beta = z$, then the original sequence of $Y(z)$ can be obtained by means of equations (9.6-8) and (9.6-9). The specific steps are as follows.

The first term on the right of equation (9.6-7) can be expressed in partial fraction form

$$Y_1(z) = \frac{z}{(z - 1)(z^2 + z + 1)} = \frac{k_1 z}{z - 1} + \frac{az^2 + bz}{z^2 + z + 1},$$

where the coefficient k_1 can be obtained by the partial fraction expansion method

$$k_1 = (z - 1)Y_1(z)|_{z=1} = \frac{1}{3}.$$

From the comparison with the coefficient method, we obtain a and b.
Since,

$$\frac{z}{(z-1)(z^2+z+1)} = \frac{\frac{1}{3}z}{z-1} + \frac{az^2+bz}{z^2+z+1} = \frac{\left(\frac{1}{3}+a\right)z^3 + \left(\frac{1}{3}+b-a\right)z^2 + \left(\frac{1}{3}-b\right)z}{(z-1)(z^2+z+1)}$$

$$(9.6\text{-}10)$$

Comparing the corresponding coefficients on both sides of equation (9.6-10), we obtain

$$a = -\frac{1}{3}, \quad b = -\frac{2}{3}.$$

Then,

$$Y_1(z) = \frac{\frac{1}{3}z}{z-1} + \frac{-\frac{1}{3}z^2 - \frac{2}{3}z}{z^2+z+1},$$

and

$$Y(z) = \frac{\frac{1}{3}z}{z-1} + \frac{-\frac{1}{3}z^2 - \frac{2}{3}z}{z^2+z+1} + \frac{z^2+3z}{z^2+z+1} = \frac{\frac{1}{3}z}{z-1} + \frac{\frac{2}{3}z^2 + \frac{7}{3}z}{z^2+z+1}.$$

If $z = -2z\cos\beta$, we have

$$\beta = \frac{2}{3}\pi$$

With the following deformation

$$\frac{\frac{2}{3}z^2 + \frac{7}{3}z}{z^2+z+1} = \frac{\frac{2}{3}\left(z^2 - z\cos\frac{2}{3}\pi\right) + \frac{4\sqrt{3}}{3}z\sin\frac{2}{3}\pi}{z^2 - 2z\cos\frac{2}{3}\pi + 1},$$

and then

$$Y(z) = \frac{\frac{1}{3}z}{z-1} + \frac{\frac{2}{3}\left(z^2 - z\cos\frac{2}{3}\pi\right)}{z^2 - 2z\cos\frac{2}{3}\pi + 1} + \frac{+\frac{4\sqrt{3}}{3}z\sin\frac{2}{3}\pi}{z^2 - 2z\cos\frac{2}{3}\pi + 1}.$$

Comparing equations (9.6-8) with (9.6-9), we obtain the inverse z transform of the equation, and so the full response is

$$y[n] = \frac{1}{3} + \frac{2}{3}\cos\frac{2\pi n}{3} + \frac{4\sqrt{3}}{3}\sin\frac{2\pi n}{3} \quad n \geq 0.$$

From the example, we can find that the difference equation analysis is similar to the s domain model analysis method for the continuous system. In the s domain model analysis method, first we change a differential equation into an algebraic equation through Laplace transform to the system model, then we obtain the Laplace transform of the system response, and finally, obtain the response in the time domain with the inverse Laplace transform. In the z domain analysis method, first we can change a difference equation into an algebraic equation by taking the z transform to the system model, then obtain the z transform of the system response, at last, obtain the response in the time domain with the inverse z transform. It can be said that the approaches are different but the results are equally satisfactory.

9.6.2 System function analysis method

As we know from the previous section,

$$\sum_{k=0}^{N} a_{N-k} z^{-k} Y_f(z) = \sum_{r=0}^{M} b_{M-r} z^{-r} F(z) .$$

Arranging the expression, we have

$$\frac{Y_f(z)}{F(z)} = \frac{\sum_{r=0}^{M} b_{M-r} z^{-r}}{\sum_{k=0}^{N} a_{N-k} z^{-k}} .$$

Similarly as for the continuous system, we can give the definition as follows.

The ratio of $Y_f(z)$ and $F(z)$ of a discrete system is called the system function or the transfer function,

$$H(z) \stackrel{\text{def}}{=} \frac{Y_f(z)}{F(z)} = \frac{\sum_{r=0}^{M} b_{M-r} z^{-r}}{\sum_{k=0}^{N} a_{N-k} z^{-k}} . \tag{9.6-11}$$

Like the conception of the system function $H(s)$ of a continuous system, $H(z)$ only depends on the structure of the system itself and the parameters of components, but has nothing to do with the excitation and the response. equation (9.6-11) shows that the system function $H(z)$ can be directly derived from the system model (the difference equation).

The impulse response $h(t)$ and the system function $H(s)$ of a continuous system constitute the Laplace transform pair. Similarly, the unit response $h[n]$ and the system function $H(z)$ of a discrete system are also the z transform pair, namely,

$$h[n] \stackrel{z}{\longleftrightarrow} H(z) . \tag{9.6-12}$$

Therefore, the $h[n]$ and $H(z)$ represent the internal characteristics of a discrete system in the time and z domains, respectively, two models are shown in ▶ Figure 9.5.

Equation (9.6-12) tells us that the system function $H(z)$ can also be obtained from the z transform of the unit response $h[n]$.

In addition to the above methods, the system function $H(z)$ can also be obtained by the transfer operator $H(E)$, that is, replacing E by z in $H(E)$, we have $H(z) = H(E)|_{E=z}$. Thus, some concepts about the transfer operator also fit for the system function, such as the characteristic equation, the characteristic root, the natural frequency, etc.

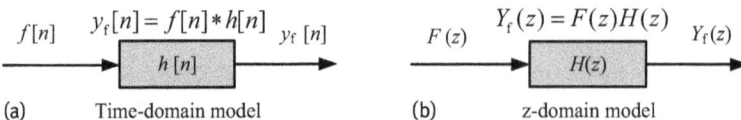

Fig. 9.5: Models of a discrete system with zero-state in time and z domains.

The system function and the unit response constitute the z transform pair and in Chapter 8 we have

$$y_f[n] = f[n] * h[n] ,$$

so, according to the convolution sum property, we have

$$y_f[n] = f[n] * h[n] \overset{z}{\longleftrightarrow} Y_f(z) = F(z)H(z) .$$

Thus, we reach the core concept of this section

$$Y_f(z) = H(z)F(z) . \tag{9.6-13}$$

This means that the z transform of the zero-state response of a discrete system is the product of the z transforms of the system function and the excitation sequence.

So far, we can give the specific steps of the analysis method from the system function:

Step 1: Find the image function $F(z)$ of the input sequence $f[n]$.

Step 2: Use the above two methods to find the system function $H(z)$.

Step 3: Use equation (9.6-13) to find the image function $Y_f(z)$ of the zero-state response.

Step 4: Use the inverse z transform to obtain the zero-state response $y_f[n]$.

Example 9.6-3. Find the unit response of the following system:

$$y[n] + \frac{1}{6}y[n-1] - \frac{1}{6}y[n-2] = f[n] - 2f[n-2] .$$

Solution. According to equation (9.6-11), the system function of this system is

$$H(z) = \frac{1 - 2z^{-2}}{1 + \frac{1}{6}z^{-1} - \frac{1}{6}z^{-2}} = \frac{z^2 - 2}{z^2 + \frac{1}{6}z - \frac{1}{6}} = \frac{\frac{3}{5}z}{z + \frac{1}{2}} + \frac{\frac{2}{5}z}{z - \frac{1}{3}} - 2z^{-2}\left(\frac{\frac{3}{5}z}{z + \frac{1}{2}} + \frac{\frac{2}{5}z}{z - \frac{1}{3}} \right) .$$

Taking the inverse z transform to the equation, we obtain the unit response

$$h[n] = \mathcal{Z}^{-1}[H(z)]$$

$$= \left[\frac{2}{5}\left(\frac{1}{3}\right)^n + \frac{3}{5}\left(-\frac{1}{2}\right)^n \right]\varepsilon[n] - 2\left[\frac{2}{5}\left(\frac{1}{3}\right)^{n-2} + \frac{3}{5}\left(-\frac{1}{2}\right)^{n-2} \right]\varepsilon[n-2]$$

$$= \left\{ \delta[n] - \frac{1}{6}\delta[n-1] - \left[\frac{34}{5}\left(\frac{1}{3}\right)^n + \frac{21}{5}\left(-\frac{1}{2}\right)^n \right]\varepsilon[n-2] \right\}\varepsilon[n] .$$

Obviously, the result from the system function method is consistent with one from the time domain method (see Example 8.5-5).

Example 9.6-4. Find the system function and the unit response of the system cascaded by two subsystems with the unit response $h_1[n] = \left(\frac{1}{2}\right)^n \varepsilon[n]$ and $h_2[n] = \left(\frac{1}{3}\right)^n \varepsilon[n]$, as shown in ▶ Figure 9.6.

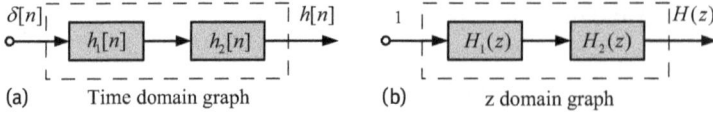

(a)   Time domain graph   (b)   z domain graph

Fig. 9.6: E9.6-4.

Solution. The output of a previous subsystem is the input of the next system, and based on the definition of the system function, we have

$$H(z) = H_1(z)H_2(z).$$

Because

$$H_1(z) = \frac{z}{z - \frac{1}{2}} \quad \text{and} \quad H_2(z) = \frac{z}{z - \frac{1}{3}},$$

the system function of this composite system is

$$H(z) = H_1(z)H_2(z) = \frac{z^2}{\left(z - \frac{1}{2}\right)\left(z - \frac{1}{3}\right)}.$$

Taking inverse z transform to the equation, the unit response is obtained as

$$h[n] = \mathcal{Z}^{-1}[H(z)] = \mathcal{Z}^{-1}\left[\frac{3z}{z - \frac{1}{2}} - \frac{2z}{z - \frac{1}{3}}\right] = \left[3\left(\frac{1}{2}\right)^n - 2\left(\frac{1}{3}\right)^n\right]\varepsilon[n].$$

Example 9.6-5. The excitation is $f_1[n] = \varepsilon[n]$, and the zero-state response is $y_{f1}[n] = 3^n\varepsilon[n]$. Find the zero-state response $y_{f2}[n]$ when the excitation is $f_2[n] = (n+1)\varepsilon[n]$.

Solution. The z transform of $f_1[n]$ and $y_{f1}[n]$ are, respectively,

$$F_1(z) = \frac{z}{z-1}, \quad |z| > 1, \qquad F_{f1}(z) = \frac{z}{z-3}, \quad |z| > 3.$$

Then

$$H(z) = \frac{Y_{f1}(z)}{F_1(z)} = \frac{z-1}{z-3}, \quad |z| > 3.$$

The z transform of $f_2[n]$ is

$$F_2(z) = \frac{z}{(z-1)^2} + \frac{z}{z-1} = \frac{z^2}{(z-1)^2}, \quad |z| > 1.$$

According to equation (9.6-13), we have

$$Y_{f2}(z) = H(z)F_2(z) = \frac{z^2}{(z-1)(z-3)} = \frac{3}{2}\frac{z}{z-3} - \frac{1}{2}\frac{z}{z-1}, \quad |z| > 3.$$

With the inverse z transform to $F_{f2}(z)$, we obtain the zero-state response

$$y_{f2}[n] = \left(\frac{3}{2}3^n - \frac{1}{2}\right)\varepsilon[n] = \frac{1}{2}\left(3^{n+1} - 1\right)\varepsilon[n].$$

Note that similarly to the continuous system, the zero-state response or the unit response of a discrete system must be multiplied by a step sequence $\varepsilon[n]$ to show that the response starts from $n = 0$; otherwise, $n \geq 0$ is required to be marked.

Excitation $f[n]$ | A LTI system (0 states) $h[n]$ | Response $y_f[n]$

$$f_1[n] = z^n$$

$$(\frac{1}{2\pi j}\frac{F(z)}{z})z^n dz$$

$$\oint_c (\frac{1}{2\pi j}\frac{F(z)}{z})z^n dz$$

$$f[n]$$

$\Longrightarrow$

$$y_{f1}[n] = H(z)z^n$$

$$H(z)(\frac{1}{2\pi j}\frac{F(z)}{z})z^n dz \quad \text{(Homogeneous)}$$

$$\oint_c H(z)(\frac{1}{2\pi j}\frac{F(z)}{z})z^n dz \quad \text{(Superposition)}$$

$$y_f[n] = \mathcal{Z}^{-1}[F(z)H(z)]$$

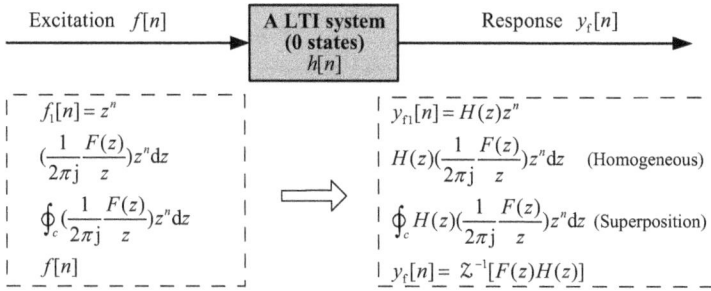

Fig. 9.7: Derivation process of the zero-state response of the system under the action of discrete signals.

9.6.3 Sequence decomposition analysis method

In the analysis of a discrete system, the z sequence z^n is as important as the imaginary exponential signal $e^{j\omega t}$ and the complex exponential signal e^{st} in the continuous system. According to equation (9.1-6), a sequence $f[n]$ can be decomposed into a linear combination of z sequences. Therefore, the analysis method of sequence decomposition can be obtained by imitating the analysis method of the continuous system.

If the input is $f[n] = z^n$, and its corresponding zero-state response is $y_{f1}[n]$, according to equation (8.5-13), we have

$$y_{f1}[n] = h[n] * f[n] = \sum_{k=0}^{n} h[k]z^{n-k} = z^n \sum_{k=0}^{n} h[k]z^{-k} = H(z)z^n . \qquad (9.6\text{-}14)$$

It states that the zero-state response to a basic sequence $f[n] = z^n$ is the product of the sequence itself and a constant coefficient independent of n. The coefficient is just the z transform of the unit response, that is, the system function $H(z)$. In this case, we can conclude that the z transform of the zero-state response of a system to any sequence is equal to the product of the z transform of the sequence and the system function,

$$Y_f(z) = H(z)F(z) . \qquad (9.6\text{-}15)$$

The procedure is similar to Figure 5.26 and is shown in ▸ Figure 9.7.

9.7 Emulation of discrete systems

A discrete system can also be emulated in the time domain or in the z domain with the basic components, and its stability can also be analyzed by the system function. Hence, in the following, readers should pay attention to the comparison with the continuous system, so as to improve their understanding of this section.

The basic operational components in a discrete system are the scalar multiplier, the shifter (delayer), and the adder, as shown in ▸ Figure 9.8.

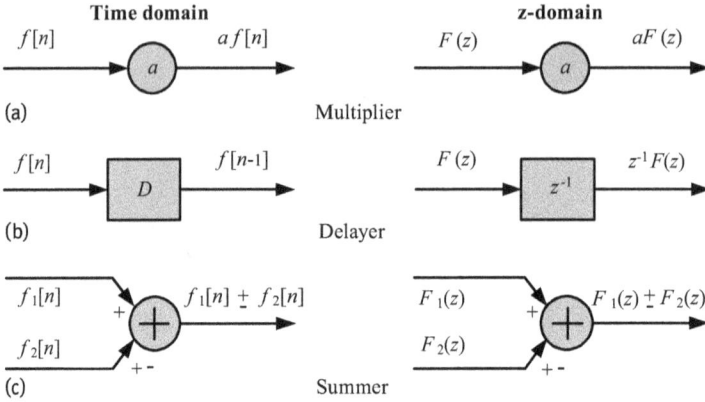

Fig. 9.8: Basic operational components in discrete systems.

In the analysis of a continuous system, we emulated a system with the transfer function $H(s)$, and the emulation method of a discrete system is analogous; it can be also divided into cascade emulation, parallel emulation and direct emulation. The main difference between the two types of system is that the emulated object changes from $H(s)$ to $H(z)$, so the character s in the emulation diagram should be replaced by

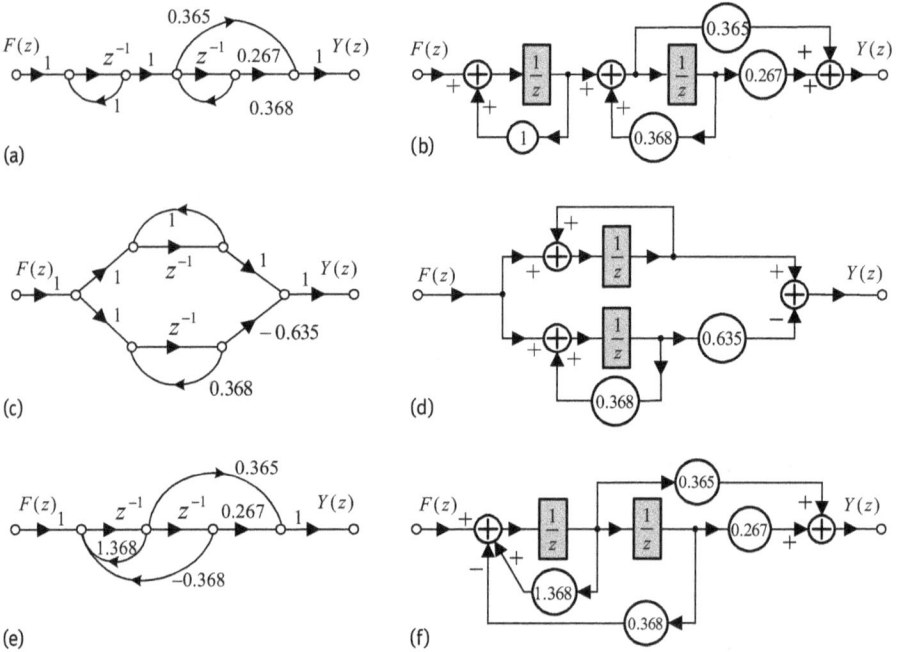

Fig. 9.9: Discrete system simulation figure of E9.7-1.

z. This means that the integrator in the continuous system should be changed into the delayer in the discrete system, but the multiplier and adder are not changed. It is necessary to explain that Mason's formula can also be used in a discrete system; the details are not discussed here.

The following example will show how to simulate a discrete system with the flow graph and the block diagram.

Example 9.7-1. The system function of a discrete system is $H(z) = \frac{0.365z+0.267}{z^2-1.368z+0.368}$. Give the direct, cascade and parallel emulation block diagram and flow graph.

Solution.

$$H(z) = \frac{0.365z + 0.267}{z^2 - 1.368z + 0.368} = \left(\frac{1}{z-1}\right)\left(\frac{0.365z + 0.267}{z - 0.368}\right) = \frac{1}{z-1} - \frac{0.635}{z - 0.368}$$

From the three different forms of the system function, the two cascade forms of the system are, respectively, as shown in ▶ Figure 9.9a and b. The parallel forms are, respectively, as shown in ▶ Figure 9.9c and d, and the direct forms are as shown in ▶ Figure 9.9e and f.

9.8 Stability analysis of discrete systems

The stability problem also exists in a discrete system; its basic concept and analysis methods are similar for the continuous system also. The stability can be defined as follows:

If the input sequence is bounded, then the corresponding output sequence is also bounded, and the discrete system is called a stable system.

For all values of n,

$$|f[n]| < M_1 < \infty \quad \rightarrow \quad |y[n]| < M_2 < \infty . \tag{9.8-1}$$

In the relation above, M_1, M_2 are finite positive numbers.

For a causal system, the necessary and sufficient condition for judging whether a system is stable is that the unit response $h[n]$ is bounded or absolutely summable, that is,

$$\sum_{n=0}^{\infty} |h[n]| < \infty . \tag{9.8-2}$$

Because the system function $H(z)$ and unit response $h[n]$ are the z transform pair,

$$H(z) = \mathcal{Z}[h(n)] = \sum_{n=0}^{\infty} h[n]z^{-n} . \tag{9.8-3}$$

Obviously, if $h[n]$ is only considered as a causal sequence, from Section 9.1 the ROC of equation (9.8-3) is outside the circle with a radius ρ. However, $h[n]$ is the unit response

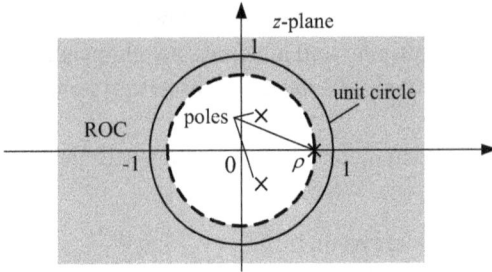

Fig. 9.10: The ROC of a causal stable system.

in fact; for a stable system, it is required to be bounded like equation (9.8-2). So, letting $z = 1$, equation (9.8-3) can be changed into

$$H(z)|_{z=1} = \sum_{n=0}^{\infty} h[n] . \tag{9.8-4}$$

Thus, equation (9.8-2) can be considered as the requirement of equation (9.8-4), that is, $H(z)|_{z=1} < \infty$, which indicates that $\rho < 1$ and that the ROC of a causal stable system is outside a circle with a radius of less than 1, written as ROC: $|z| > \rho |_{\rho<1}$. In other words, the necessary and sufficient condition to ensure an LTI causal system is BIBO stable is that the ROC of the system function $H(z)$ should include the unit circle; the ROC is shown in ▶ Figure 9.10. Note: The convergence radius ρ cannot equal $|z|$, because the ROC cannot include its own boundary.

In Chapter 7, we have learned that the stability of the continuous system can be determined by the distribution of poles. For first-order poles, the related conclusions are as follows.

(1) If the poles are all located in the left half-plane of the s domain, the response is convergent and the system is stable.

(2) As long as one pole is located in the right half-plane of the s domain, the response is divergent and the system is instable.

(3) If the poles are located on the $j\omega$ axis (imaginary axis), the response is oscillatory and the system is boundary stable.

These conclusions can be easily transferred to discrete systems. Because the imaginary axis in the s domain can be mapped to the unit circle in the z domain, the left half-plane in the s domain can be mapped to the inside the unit circle in the z domain, and the right half-plane in the s domain can be mapped to the outside the unit circle in the z domain. Thus, we will discuss the relationships between the distribution of poles of $H(z)$ and the change of $h[n]$.

We rewrite the system function $H(z)$ as follows:

$$H(z) = \frac{\sum_{r=0}^{M} b_{M-r} z^{-r}}{\sum_{k=0}^{N} a_{N-k} z^{-k}} = H_0 \frac{\prod_{r=1}^{M} (z - z_r)}{\prod_{k=1}^{N} (z - \lambda_k)} , \tag{9.8-5}$$

where z_r and λ_k are the zeros and the poles. Letting all poles be of first order, using partial fraction expansion and letting $\lambda_0 = 0$, we have

$$H(z) = \sum_{k=0}^{N} \frac{A_k z}{z - \lambda_k} = A_0 + \sum_{k=1}^{N} \frac{A_k z}{z - \lambda_k} . \tag{9.8-6}$$

The inverse z transform of the equation is

$$h[n] = \mathcal{Z}^{-1}[H(z)] = A_0 \delta[n] + \sum_{k=1}^{N} A_k \lambda_k^n \varepsilon[n] . \tag{9.8-7}$$

It can be seen that the form of the unit response $h[n]$ can be totally determined by the positions of the poles λ_k. This way, the relationships between the pole positions and the stability of a discrete system can be expressed as

(1) If $|\lambda_k| < 1$, the poles are located inside the unit circle, because $\lim_{n \to \infty} A_k \lambda_k^n = 0$, and the amplitudes of corresponding terms in $h[n]$ are attenuated, and the system is stable. The ROC is the outside a circle with a radius of less than 1.

(2) If $|\lambda_k| > 1$, the poles are located outside the unit circle, because $\lim_{n \to \infty} A_k \lambda_k^n \to \infty$, and the amplitudes of corresponding terms in $h[n]$ grow, and the system is unstable.

(3) If $|\lambda_k| = 1$, the poles are all located on the unit circle, because $\lim_{n \to \infty} A_k \lambda_k^n = A_k$, the amplitudes of corresponding terms in $h[n]$ remain constant, and the system is marginally stable or critically stable.

▶ Figure 9.11 shows these relationships. It is emphasized that only the situation of first-order poles is discussed here; for a treatment of higher order poles the reader is referred to other books.

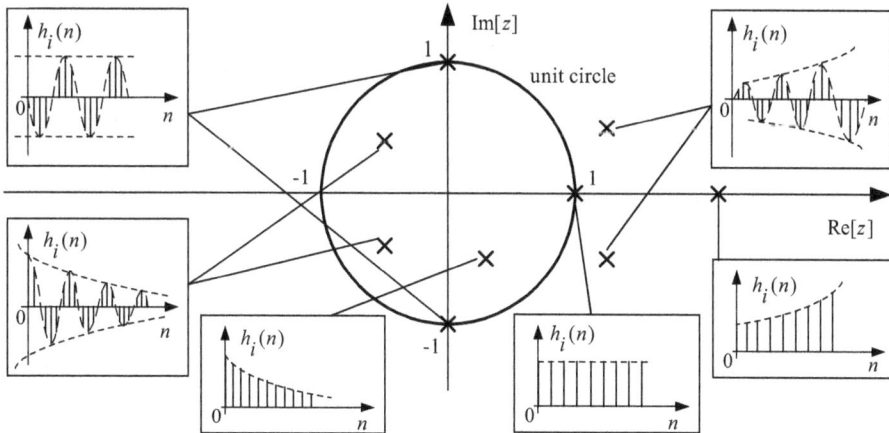

Fig. 9.11: Relations between first-order ploes of $H(z)$ and waveforms of $h[n]$.

In addition, similarly to the Routh–Hurwitz criterion in a continuous system, there is also the E. I. Jury criterion to judge the system stability of a discrete system. If the system function is $H(z) = \frac{B(z)}{A(z)}$ and the characteristic polynomial is $A(z) = a_2 z^2 + a_1 z + a_0$, then the sufficient and necessary condition of which this second-order system is stable is

$$\begin{cases} A(1) > 0 \\ A(-1) > 0 \\ a_2 > |a_0| \end{cases} . \qquad (9.8\text{-}8)$$

Example 9.8-1. The system function of a system is $H(z) = \frac{z+1}{z^2+(2+k)z+0.5}$. Find values of k that can make the system stable.

Solution. The characteristic polynomial is

$$A(z) = z^2 + (2 + k)z + 0.5 .$$

From the E. I. Jury criterion, there should be

$$A(1) = 1 + (2 + k) + 0.5 > 0$$
$$A(-1) = 1 - (2 + k) + 0.5 > 0 .$$

This can be obtained by

$$k > -3.5 \quad \text{and} \quad k < -0.5.$$

Namely, when $-3.5 < k < -0.5$, the system is stable.

The E. I. Jury criterion will not be discussed in more detail in this book.

9.9 Analysis methods of discrete systems in the frequency domain

Similarly to the continuous system, the discrete system also uses the analysis method in the frequency domain, and the main analysis tools are the discrete time Fourier series and the discrete time Fourier transform or, simply, DTFS and DTFT. Correspondingly, the continuous-time Fourier series and the Fourier transform introduced in Chapters 4 and 5 can be abbreviated as CTFS and CTFT.

9.9.1 Discrete-time fourier series

Let the discrete independent variables in the time domain and frequency domain be, respectively, n and Ω, $f[n]$ be a discrete periodic sequence with period N, so DTFS is defined as

$$f[n] = \sum_{k=0}^{N-1} F(k\Omega_0) e^{jk\Omega_0 n} \quad n = 0, 1, \ldots, N-1 , \qquad (9.9\text{-}1)$$

where

$$F(k\Omega_0) = \frac{1}{N} \sum_{n=0}^{N-1} f[n]e^{-jk\Omega_0 n} \quad k = 0, 1, \ldots, N-1 ; \tag{9.9-2}$$

$F(k\Omega_0)$ is also called the spectrum of the sequence $f[n]$.

The above two expressions can be related by

$$f[n] \xleftrightarrow{\mathcal{DTFS}} F(k\Omega_0) . \tag{9.9-3}$$

The CTFS and the DTFS have the following corresponding relationship:

$$\text{CTFS}|_{\omega\to\Omega, \omega_0\to\Omega_0, t\to n, T\to N} \to \text{DTFS} \tag{9.9-4}$$

Obviously, the DTFS is similar to the CTFS in concept and meaning, and can decompose a discrete periodic signal into an algebraic sum of discrete imaginary exponential signals, and it is used to analyze the responses of a discrete system to periodic sequences; the analysis method is also similar to that of the CTFS.

9.9.2 Discrete time fourier transform

The Fourier transform $F(\Omega)$ of a nonperiodic sequence $f[n]$ is defined as

$$F(\Omega) = \sum_{n=-\infty}^{+\infty} f[n]e^{-j\Omega n} . \tag{9.9-5}$$

The inverse Fourier transform is defined as

$$f[n] = \frac{1}{2\pi} \int_{(2\pi)} F(\Omega)e^{j\Omega n}\,d\Omega . \tag{9.9-6}$$

The $f[n]$ and $F(\Omega)$ are the Fourier transform pair,

$$f[n] \xleftrightarrow{\mathcal{DTFT}} F(\Omega) \tag{9.9-7}$$

Like the concept of the continuous signal, $F(\Omega)$ is also known as the frequency spectrum function of the sequence $f[n]$. As a result, it can be written as

$$F(\Omega) = |F(\Omega)|\, e^{j\varphi(\Omega)} , \tag{9.9-8}$$

where $|F(\Omega)|$ is called the amplitude spectrum of $f[n]$ (amplitude frequency characteristic), and $\varphi(\Omega)$ is called the phase spectrum (phase frequency characteristic). Note that $F(\Omega)$ can be also written as $F\left(e^{j\Omega}\right)$.

Besides some similar features to the CTFT, such as linearity, time shifting, frequency shifting and convolution, the DTFT has also the following two features:

(1) Periodicity. The discrete Fourier transform $F(\Omega)$ is a periodic function with a period 2π, that is, $F(\Omega + 2\pi) = F(\Omega)$.

(2) The magnitude $|F(\Omega)|$ of $F(\Omega)$ is an even function of Ω, and the phase $\varphi(\Omega)$ is an odd function of Ω. The real component $\text{Re}[F(\Omega)]$ of $F(\Omega)$ is an even function of Ω, and the imaginary component $\text{Im}[F(\Omega)]$ is an odd function of Ω.

9.9.3 Analysis method of discrete systems with fourier transform

Let the Fourier transform of the excitation for a discrete LTI system be $F(\Omega)$, and the Fourier transform of the zero-state response be $Y(\Omega)$, so the system function $H(\Omega)$ in the frequency domain is defined as

$$H(\Omega) = \frac{Y(\Omega)}{F(\Omega)} . \tag{9.9-9}$$

Obviously, the Fourier transform $Y(\Omega)$ of the zero-state response for a discrete LTI system can be found by following equation:

$$Y(\Omega) = F(\Omega)H(\Omega) . \tag{9.9-10}$$

This equation corresponds to the expression $y[n] = f[n] * h[n]$ in the time domain.

Example 9.9-1. The unit response $h[n] = \left(\frac{1}{2}\right)^2 \varepsilon[n]$ and the input $f[n] = \left(\frac{1}{4}\right)^2 \varepsilon[n]$ of an LTI discrete system are known. Find the spectrum of the response.

Solution. The system function in the frequency domain and the spectrum of the input are

$$H(\Omega) = \mathcal{F}[h[n]] = \sum_{n=-\infty}^{\infty} \left(\frac{1}{2}\right)^n e^{-j\Omega n} = \frac{1}{1 - 0.5e^{-j\Omega}},$$

$$F(\Omega) = \mathcal{F}[f[n]] = \sum_{n=-\infty}^{\infty} \left(\frac{1}{4}\right)^n e^{-j\Omega n} = \frac{1}{1 - 0.25e^{-j\Omega}}.$$

So, from equation (9.9-10) we can obtain

$$Y(\Omega) = F(\Omega)H(\Omega) = \frac{1}{1 - 0.5e^{-j\Omega}} \frac{1}{1 - 0.25e^{-j\Omega}} = \frac{1}{(1 - 0.5e^{-j\Omega})(1 - 0.25e^{-j\Omega})} .$$

9.9.4 Frequency characteristic of the discrete system

When the ROC of the system function $H(z)$ of a discrete system includes the unit circle and z changes along the unit circle, that is, $H(z)|_{z=e^{j\Omega}} = H\left(e^{j\Omega}\right)$, then $H(z)$ represents the changes of the amplitude and phase with Ω; $H\left(e^{j\Omega}\right)$ is called the frequency response of the discrete system. Note that $H\left(e^{j\Omega}\right)$ can also be written as $H(\Omega)$.

As with the continuous system, the frequency characteristic of a discrete system is also classified as the amplitude frequency characteristic $\left|H\left(e^{j\Omega}\right)\right|$ ($|H(\Omega)|$) and the phase frequency characteristic $\varphi(\Omega)$, that is,

$$H\left(e^{j\Omega}\right) = \sum_{n=0}^{\infty} h[n]e^{-j\Omega n} = H(z)|_{z=e^{j\Omega}} = \left|H\left(e^{j\Omega}\right)\right| e^{j\varphi(\Omega)} . \tag{9.9-11}$$

Obviously, the unit response $h[n]$ and the frequency characteristic $H\left(e^{j\Omega}\right)$ are the Fourier transform pair.

When the input sequence is an arbitrary frequency sine sequence or an imaginary exponential sequence,

$$f[n] = Ae^{j\Omega n} \quad -\infty < n < \infty .$$

Imitating the method in Section 5.6.3, we obtain

$$y[n] = H\left(e^{j\Omega}\right)f[n] . \tag{9.9-12}$$

For example, if $f[n] = A \sin \Omega n$, the corresponding sine steady state response is

$$y[n] = A\left|H\left(e^{j\Omega}\right)\right| \sin[\Omega n + \varphi(\Omega)] . \tag{9.9-13}$$

So far, some points for the analysis of discrete time LTI systems are as follows.
(1) The steady state response of a stable causal system to a sinusoidal sequence is still a sine sequence with the same envelope frequency.
(2) The amplitude and phase frequency characteristics $\left|H\left(e^{j\Omega}\right)\right|$ and $\varphi(\Omega)$ respectively represent the relative changes between the output sequence and the input sequence on the amplitude and phase. In general, they are a function of the envelope frequency Ω.
(3) The amplitude frequency characteristic $\left|H\left(e^{j\Omega}\right)\right|$ is an even function of Ω, while the phase frequency characteristic $\varphi(\Omega)$ is an odd function of Ω.
(4) Since $e^{j\Omega}$ is periodical, $H\left(e^{j\Omega}\right)$ is also periodical with 2π period. This is a feature of discrete system frequency response.

Note that when the sampling interval is $T = 1$, the digital frequency $\Omega = \omega T$ will become the angular frequency ω, that is, $\Omega \overset{T=1}{=} \omega$.

Example 9.9-2. The difference equation of a discrete system is as follows

$$y[n] - ay[n-1] = f[n] \quad (0 < a < 1) .$$

(1) Draw the emulation block diagram in the time domain.
(2) Obtain $H(z)$ and the pole-zero diagram;
(3) if the ROC of $H(z)$ is $|z| > a$, find the unit response $h[n]$.
(4) Find the amplitude and phase frequency characteristics of the system and roughly draw their curves.
(5) If $f[n] = 5 + 12 \sin \frac{\pi}{2}n - 20 \sin\left(\pi n + \frac{\pi}{4}\right)$, find the response $y[n]$, where $a = 0.9$.

Solution. (1) $y[n] = f[n] + ay[n-1]$; its emulation block diagram in the time domain is shown in ▶ Figure 9.12a.
(2) $H(z) = \frac{z}{z-a}$, the zero and pole are, respectively, $z_1 = 0$ and $p_1 = a$; the pole-zero diagram is shown in ▶ Figure 9.12b.
(3) Because the ROC of $H(z)$ is $|z| > a$, from Table 9.1 we know that the unit response is a right-sided sequence, $h[n] = a^n \varepsilon[n]$.

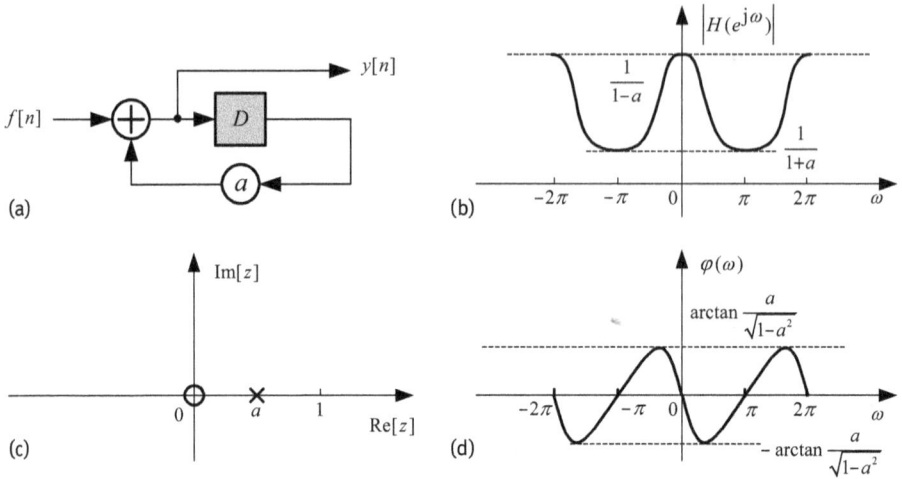

Fig. 9.12: E9.9-1.

(4) Because the ROC of $H(z)$ is $|z| > a$, and $0 < a < 1$. Set $T = 1$, $\Omega = \omega T = \omega$, and so the frequency characteristic is

$$H(e^{j\Omega}) = H\left(e^{j\omega}\right) = H(z)|_{z=e^{j\omega}} = \frac{1}{1 - a\cos\omega + ja\sin\omega}.$$

We have

$$\left|H\left(e^{j\omega}\right)\right| = \frac{1}{\sqrt{1 + a^2 - 2a\cos\omega}}$$

$$\varphi(\omega) = -\arctan\frac{a\sin\omega}{1 - a\cos\omega}.$$

The frequency characteristic curves for the amplitude and phase can be pictured by the point depiction method shown in ▶ Figure 9.12c and d. It can be seen that the system is a low pass filter. Moreover, it can be proved that the system would become a high pass filter if $-1 < a < 0$.

(5) The frequency components included by the excitation are terms $\Omega = 0, \frac{\pi}{2}, \pi$, so we have

$$|H(0)| = \frac{1}{1 - a} = 10, \quad \varphi(0) = 0$$

$$\left|H\left(\frac{\pi}{2}\right)\right| = \frac{1}{\sqrt{1 + a^2}} = \frac{1}{\sqrt{1.81}} = 0.74, \quad \varphi\left(\frac{\pi}{2}\right) = -\arctan(a) = -42°$$

$$|H(\pi)| = \frac{1}{1 + a} = \frac{1}{1.9} = 0.53, \quad \varphi(\pi) = -\arctan(0) = 0$$

By equation (9.9-13) and the linear property of the system, the response is

$$y[n] = 5\,|H(0)| + 12\left|H\left(\frac{\pi}{2}\right)\right|\sin\left[\frac{\pi}{2}n + \varphi\left(\frac{\pi}{2}\right)\right] - 20\,|H(\pi)|\sin\left[\pi n + \frac{\pi}{4} + \varphi(\pi)\right]$$

$$= 50 + 8.88\sin\left(\frac{\pi}{2}n - 42°\right) - 10.6\sin\left(\pi n + \frac{\pi}{4}\right) \quad -\infty < n < \infty$$

For details about the discrete systems and signal analysis in the frequency domain, the interested reader is referred to other references.

9.10 Concept comparisons between discrete systems and continuous systems

Based on the study of Chapter 8 and Chapter 9, the similarities in concept and analyses between discrete and continuous signals and discrete and continuous systems can be seen; there are many duality relations between them. Understanding and mastering these differences and similarities will greatly help us to have the relevant knowledge

Tab. 9.5: Concept comparison between continuous and discrete signals and systems.

Continuous signals and systems	Discrete signals and systems
Signal $f(t)$	Sequence $f[n]$
Differential	Difference
Integral	Cumulative sum
Convolution	Discrete convolution
Unit step signal $\varepsilon(t)$	Unit step sequence $\varepsilon[n]$
Unit impulse signal $\delta(t)$	Unit impulse sequence $\delta[n]$
The s domain	The z domain
The Laplace transform	The z transform
Variable $s = \sigma + j\omega$	Variable $z = re^{j\Omega}$
Basic signal e^{st}	Basic signal z^n
Basic signal $e^{j\omega t}$	Basic signal $e^{j\Omega}$
Differential equation	Difference equation
Differential operator p	Advance operator E
Integral operator $1/p$	Lag operator $1/E$
Transport operator $H(p)$	Transport operator $H(E)$
Impulse response $h(t)$	Impulse response $h[n]$
System function $H(s)$	System function $H(z)$
Energy storage component	Delay component
Integrator $1/s$	Delayer $1/z$
Imaginary axis $j\omega$	The unit circle $r = 1$
Starting conditions $\left\{y^{(n)}(0_-)\right\}$	Starting conditions $\{y[-n]\}\ n = 1, 2, \ldots, N$
Initial conditions $\left\{y^{(n)}(0_+)\right\}$	Initial conditions $\{y[n]\}\ n = 0, 1, 2, \ldots, N - 1$

of discrete signals and systems. Readers are expected to think about the relationship between them to improve their study efficiency.

For convenience, some main concepts related to continuous and discrete signals and systems are listed in Table 9.5. From careful observation and analysis of the characteristics, we find that they are connected by two bridges: the sampling theorem and the transformation between the s plane and the z plane.

9.11 Solved questions

Question 9-1. The z transform of a discrete sequence is $F(z) = \frac{4}{z^2(2z-1)}$, and the ROC is $|z| > 0.5$. Solve the original sequence $f[n]$.

Solution. Rewriting $F(z)$ as

$$F(z) = \frac{4}{z^2(2z-1)} = \frac{2}{z^2\left(z-\frac{1}{2}\right)} = \frac{8}{z-\frac{1}{2}} - 8z^{-1} - 4z^{-2},$$

in the ROC $|z| > 0.5$, we have

$$\frac{8}{z-\frac{1}{2}} \xleftrightarrow{z} 8\left(\frac{1}{2}\right)^{n-1}\varepsilon[n-1], \quad z^{-1} \xleftrightarrow{z} \delta[n-1] \quad \text{and} \quad z^{-2} \xleftrightarrow{z} \delta[n-2].$$

With the inverse z transform applied to $F(z)$, the original sequence $f[n]$ can be obtained,

$$f[n] = 8\left(\frac{1}{2}\right)^{n-1}\varepsilon[n-1] - 8\delta[n-1] - 4\delta[n-2].$$

Question 9-2. The difference equation of a discrete system is $y[n] + \frac{1}{2}y[n-1] = x[n] + x[n-1]$, the input sequence is $x[n] = \left(\frac{1}{2}\right)^n \varepsilon[n]$ and $y[-1] = 2$. Find the output sequence $y[n]$.

Solution. With the z transform applied to both sides of the difference equation,

$$Y(z) + \frac{1}{2}z^{-1}Y(z) + \frac{1}{2}y(-1) = X(z) + z^{-1}X(z).$$

Substituting the z transform $X(z) = \frac{1}{1-\frac{1}{2}z^{-1}}$ of the input signal $x[n] = \left(\frac{1}{2}\right)^n \varepsilon[n]$ into the above equation and using the starting condition $y[-1] = 2$, we obtain the output signal in the z domain

$$Y(z) = \frac{1+z^{-1}}{1+\frac{1}{2}z^{-1}}X(z) - \frac{1}{1+\frac{1}{2}z^{-1}} = \frac{1+z^{-1}}{\left(1+\frac{1}{2}z^{-1}\right)\left(1-\frac{1}{2}z^{-1}\right)} - \frac{1}{1+\frac{1}{2}z^{-1}}$$

$$= \frac{3/2}{1-\frac{1}{2}z^{-1}} - \frac{1/2}{1+\frac{1}{2}z^{-1}}.$$

Applying the inverse z transform to the equation, we obtain the output signal,

$$y[n] = \left[\frac{3}{2}\left(\frac{1}{2}\right)^n - \frac{1}{2}\left(-\frac{1}{2}\right)^n\right]\varepsilon[n].$$

Question 9-3. The difference equation of a discrete system is

$$y[n] - y[n-1] - \frac{3}{4}y[n-2] = f[n-1].$$

(1) Find the system function $H(z)$.
(2) Find three possible options for the unit response $h[n]$ and discuss the stability and causality of each one.

Solution. (1) With the z transform on both sides of the difference equation, we have

$$Y(z) - Y(z)z^{-1} - \frac{3}{4}Y(z)z^{-2} = F(z)z^{-1}.$$

The definition of the system function $H(z) = \frac{Y(z)}{F(z)}$ yields

$$H(z) = \frac{z}{z^2 - z - \frac{3}{4}} = \frac{z}{\left(z + \frac{1}{2}\right)\left(z - \frac{3}{2}\right)} = \frac{-\frac{1}{2}z}{z + \frac{1}{2}} + \frac{\frac{1}{2}z}{z - \frac{3}{2}}.$$

There are two poles $p_1 = -\frac{1}{2}, p_2 = \frac{3}{2}$ and one zero $z = 0$ in $H(z)$. To obtain $h[n]$ by the inverse z transform of $H(z)$ in the corresponding ROC, we need to discuss the ROCs of the system function. When ROC is $|z| > \frac{3}{2}$, $h[n] = [-\frac{1}{2}(-\frac{1}{2})^n + \frac{1}{2}(\frac{3}{2})^n]\varepsilon[n]$ is a right-sided sequence. Here, the system is a causal instable system because the ROC does not include the unit cycle. When ROC is $\frac{1}{2} < |z| < \frac{3}{2}$, $h[n] = -\frac{1}{2}\left(-\frac{1}{2}\right)^n \varepsilon[n] - \left(\frac{3}{2}\right)^n \varepsilon[-n-1]$ is a bilateral sequence. Here, the system is a noncausal stable system because the ROC includes the unit cycle. When ROC is $|z| < \frac{1}{2}$, $h[n] = \frac{1}{2}[(-\frac{1}{2})^n - (\frac{3}{2})^n]\varepsilon[-n-1]$ is a left-sided sequence. At this time, the system is a noncausal instable system because the ROC does not include the unit cycle.

Question 9-4. A discrete causal LTI system is described by a difference equation $y[n] - y[n-1] - 6y[n-2] = f[n-1]$.
(1) Find the system function and its ROC.
(2) Find the unit impulse response $h[n]$ of the system.
(3) When $f[n] = (-3)^n$, $-\infty < n < \infty$. Find the output $y[n]$.

Solution. (1) Applying the z transform to this equation, we have

$$Y_f(z) = \frac{1}{1 - z^{-1} - 6z^{-2}}z^{-1}F(z).$$

Arranging the equation, the system function is of the form

$$H(z) = \frac{Y_f(z)}{F(z)} = \frac{z^{-1}}{1 - z^{-1} - 6z^{-2}} = \frac{z}{z^2 - z - 6} = \frac{z}{(z+2)(z-3)}.$$

From the equation, poles of $H(z)$ are, respectively, $z_1 = -2$ and $z_2 = 3$. Considering that the system is causal, the ROC is $|z| > 3$.

(2) According to the partial fraction expansion method, we have

$$H(z) = \frac{z}{(z+2)(z-3)} = \frac{1}{5}\left(\frac{2}{z+2} + \frac{3}{z-3}\right).$$

Applying the inverse z transform to the equation, we obtain the unit impulse response,

$$h[n] = \frac{1}{5}\left[2(-2)^{n-1} + 3(3)^{n-1}\right]\varepsilon[n-1] = \frac{1}{5}\left[(3)^n - (-2)^n\right]\varepsilon[n-1].$$

(3) Because the definition of the z transform is $H(z) = \sum_{n=-\infty}^{\infty} h[n]z^{-n}$, we can find $y[n]$ by means of the convolution sum,

$$y[n] = f[n] * h[n] = \sum_{k=-\infty}^{\infty} f[n-k]h[k] = \sum_{k=-\infty}^{\infty}(-3)^{n-k}h[k] = (-3)^n \sum_{k=-\infty}^{\infty}(-3)^{-k}h[k]$$

$$= (-3)^n H(z)\big|_{z=-3} = (-3)^n \left(\frac{z}{(z+2)(z-3)}\right)\bigg|_{z=-3} = -\frac{1}{2}(-3)^n$$

9.12 Learning tips

The z domain analysis is as important as the time domain analysis, so the reader is advised to pay attention to the following:
(1) The similarities and differences between the z transform and the Laplace transform.
(2) The solutions for the inverse z transform.
(3) The similarities and differences of the system functions in the real frequency domain, the complex frequency domain and the z domain.
(4) The similarities and differences between the analysis methods in the z domain and the s domain.

9.13 Problems

Problem 9-1. Find the unilateral z transforms for the following sequences using the definition, show the regions of convergence, and point out the zeros and poles of $F(z)$.

(1) $f[n] = [\underset{\uparrow}{1}, -1, 1, -1, \ldots]$

(2) $f[n] = [\underset{\uparrow}{0}, 1, 0, 1, \ldots]$

(3) $f[n] = \delta[n-N]$

(4) $f[n] = \delta[n+N]$

(5) $f[n] = \left(\frac{1}{2}\right)^n \varepsilon[n-1]$

(6) $f[n] = \left(\frac{1}{2}\right)^n \varepsilon[-n]$

(7) $f[n] = \left(\frac{1}{2}\right)^{|n|}$

(8) $f[n] = \left(\frac{1}{4}\right)^n \varepsilon[n] - \left(\frac{2}{3}\right)^n \varepsilon[n]$

Problem 9-2. Find the bilateral z transforms of the following sequences and indicate the ROC for each of them.

(1) $f[n] = \left(\frac{1}{2}\right)^n \varepsilon[-n]$

(2) $f[n] = \left(\frac{1}{2}\right)^{|n|}$

Problem 9-3. Find the z transforms of the following sequences using the properties of z transform.

(1) $f[n] = \varepsilon[n] - \varepsilon[n-8]$

(2) $f[n] = \cos\left(\frac{n\pi}{2}\right) \cdot \varepsilon[n]$

(3) $f[n] = \left(\frac{1}{2}\right)^n \cos\left(\frac{n\pi}{2}\right) \cdot \varepsilon[n]$

(4) $f[n] = n\varepsilon[n]$

(5) $f[n] = n(n-1)\varepsilon[n]$

(6) $f[n] = n\alpha^n \varepsilon[n]$

(7) $f[n] = (n-3)\varepsilon[n-2]$

(8) $f[n] = (n-2)\varepsilon[n-2]$

(9) $f[n] = (-n-2)\varepsilon[-n]$

(10) $f[n] = \begin{cases} n & 0 \le n \le 4 \\ 8-n & 5 \le n \le 8 \\ 0 & \text{other} \end{cases}$

Problem 9-4. Based on the series $e^x = 1 + x + \frac{x^2}{2!} + \frac{x^3}{3!} + \cdots + \frac{x^n}{n!} + \cdots$, find the image function $F_1(z) = e^{-\frac{a}{z}}$ (the whole z plane except $z = 0$) and $F_2(z) = e^z(|z| < \infty)$, and the corresponding sequences $f_1[n]$ and $f_2[n]$.

Problem 9-5. Using the power series expansion method, find the corresponding original functions (the first four terms) for the following image functions:

(1) $F_1(z) = \frac{2z^2 - 0.5z}{z^2 - 0.5z - 0.5}$;

(2) $F_2(z) = \frac{z}{z^2 - 1}$;

(3) $F_3(z) = \frac{z^3 + 2z^2 + 1}{z^3 - 1.5z^2 + 0.5z}$

Problem 9-6. Using the partial fraction expansion method, find the corresponding original sequences for the following image functions:

(1) $F(z) = \frac{10z^2}{z^2 - 1}$ $(|z| > 1)$

(2) $F(z) = \frac{2z^2 - 3z + 1}{z^2 - 4z - 5}$ $(|z| > 5)$

(3) $F(z) = \frac{z+1}{(z-1)^2}$ $(|z| > 1)$

(4) $F(z) = \frac{z^2 - 1}{(z+\frac{1}{2})(z-\frac{1}{3})}$ $(|z| > \frac{1}{2})$

Problem 9-7. Find the inverse z transforms for the following image functions:

(1) $F(z) = 7z^{-1} + 3z^{-2} - 8z^{-10}$

(2) $F(z) = 2z + 3 + 4z^{-1}$

(3) $F(z) = \frac{z-5}{z+2}$ $(|z| > 2)$

(4) $F(z) = \frac{z^{-1}}{(1 - 6z^{-1})^2}$ $(|z| > 6)$

(5) $F(z) = \log(1 - 2z)$

(6) $F(z) = \frac{1}{z^2 + 1}$ $(|z| > 1)$

(7) $F(z) = \frac{z^2}{z^2 - \sqrt{3}z + 1}$ $(|z| > 1)$

(8) $F(z) = \frac{z^2}{z^2 + \sqrt{2}z + 1}$ $(|z| > 1)$

Problem 9-8. If $F(z) = \frac{z+2}{2z^2 - 7z + 3}$, find the corresponding sequences with different three kinds of convergence regions in the following:

(1) $|z| > 3$

(2) $|z| < \frac{1}{2}$

(3) $\frac{1}{2} < |z| < 3$

Problem 9-9. Using the convolution sum theorem, solve the following convolution sums:

(1) $a^n \varepsilon[n] * \delta[n-2]$

(2) $a^n \varepsilon[n] * \varepsilon[n+1]$

(3) $a^n \varepsilon[n] * b^n \varepsilon[n]$

(4) $\varepsilon[n-2] * \varepsilon[n-2]$

(5) $n\varepsilon[n] * \varepsilon[n]$

(6) $a^n \varepsilon[n] * b^n \varepsilon[-n]$

Problem 9-10. Using the z transform, find the full response for the following difference equations:

(1) $y[n] + 3y[n-1] + 2y[n-2] = \varepsilon[n]$, $y[-1] = 0$, $y[-2] = \frac{1}{2}$

(2) $y[n] + 2y[n-1] + y[n-2] = \frac{4}{3} \cdot 3^n \varepsilon[n]$, $y[-1] = 0$, $y[0] = \frac{4}{3}$

(3) $y[n+2] + y[n+1] + y[n] = \varepsilon[n]$, $y[0] = 1$, $y[1] = 2$

Problem 9-11. A first-order discrete system is shown in ▶ Figure P9-11. Find the zero-state responses to the unit step sequence $\varepsilon[n]$ and the complex exponential sequence $e^{j\omega n}\varepsilon[n]$. Note: $0 < a < 1$.

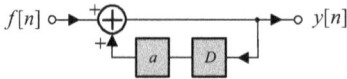

Fig. P9-11

Problem 9-12. Find the system functions for the following systems and determine the stability for each of them.

(1) A system with block diagram shown as ▶ Figure P9-12a

(2) A system with flow diagram shown as ▶ Figure P9-12b

(3) A system with unit response shown as ▶ Figure P9-12c

(4) A system with the difference equation $y[n+2] + 2y[n+1] + 2y[n] = f[n+1] + 2f[n]$

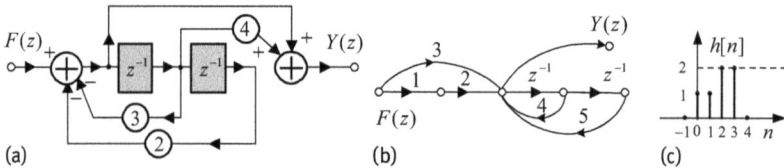

(a)

(b)

(c)

Fig. P9-12

Problem 9-13. Two system functions are as follows. Give the emulation block diagrams and flow diagrams in their direct, parallel and series forms:

(1) $H(z) = \frac{3+3.6z^{-1}+0.6z^{-2}}{1+0.1z^{-1}-0.2z^{-2}}$; (2) $H(z) = \frac{z^2}{(z+0.5)^3}$

Problem 9-14. The emulation block diagram of a discrete system is shown in ▶ Figure P9-14.

(1) Find the system function $H(z)$.

(2) Find the range of values of K when the system is stable.

(3) Find the unit response $h[n]$ when the system is critically stable.

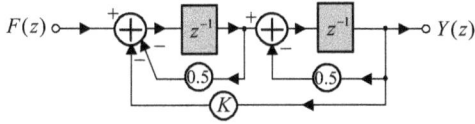

$F(z) \circ$ $\longrightarrow$ $Y(z)$

Fig. P9-14

Problem 9-15. A system is shown in ▶ Figure P9-15.
(1) Find the system function $H(z)$.
(2) Judge the stability of the system.
(3) Write difference equation of the system.
(4) Find the unit response $h[n]$;
(5) If an input $f[n] = \varepsilon[n]$, and $y_x[0] = y_x[1] = 1$, find the zero-input response, the zero-state response and the full response.

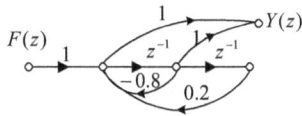

Fig. P9-15

10 State space analysis of systems

Questions: The system analysis methods discussed in the previous chapters are all based on mathematical input-output models of a system. These models are used to study the transformation characteristics of a given system to any signal. Can the system response be influenced by the changes of internal parameters or structure of a system? If so, what is the impact on the system response? In addition, how should we analyze complex systems with multiple inputs and outputs?

Solution: Seek parameters that can reflect the internal characteristics of a system; associate the excitations and responses with these parameters; introduce matrix tools to solve the above problems.

Results: State space analysis method.

10.1 State space description of a system

1. Description based on input–output ports

As stated in the previous chapters, in order to analyze a SISO–LTI system, we must determine an input-output mathematical model, which is usually an nth-order linear differential (difference) equation. Once this model is built, we will no longer care about the internal changes of the system and only need to consider how the output of the system changes with the input. This description method, which only focuses on the changes of the physical quantities on system ports with time or frequency variables, is called the external description method or the port description method.

With an increasing number of ports and structure complexity, it is inconvenient or difficult to analyze an LTI system by using the external method. On the other hand, with the development of technology and changes in the requirements of people, in the analysis of some control systems we can no longer be content with just the study of the input-output relationship. We also want to know the change regulars for some internal parameters of the system, so that we can achieve optimal control by designing and adjusting these parameters. Obviously, now the external method corresponds to "ability not equal to one's ambition". Thus, a second system description method emerges at a historic moment, which is called the state space description method. The schematic diagrams of two description methods for a SISO system are shown in ▶ Figure 10.1.

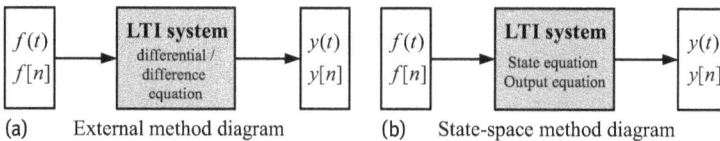

(a) External method diagram   (b) State-space method diagram

Fig. 10.1: Two description diagrams of SISO system.

https://doi.org/10.1515/9783110541205-003

2. Description based on the state space

The state space representation of a system is a method in which the differential equation describing the output-input relationship of a system is not given directly, but first a set of auxiliary variables (state variables) are appropriately selected in the system, then a set of first-order differential equations relating to these variables and the system inputs are found and are called the state equations, and finally, a set of algebraic expressions relating to these variables and the inputs and the outputs of the system are found and are called the output equations, so that the work relating to the system inputs, the state variables, and the system outputs has been finished. The procedure describing a system with the state equations and the output equations is called the state space description method. The state equations and output equations can be collectively referred to as the state model of the system. The approach analyzing a system from the state space description is called the state space analysis method. Note: The state space description method is only suitable for dynamic systems.

Next, we will introduce several common terms in the analysis method. Note that concepts such as "the state" and "the initial state" here are the continuation and supplement of related concepts in Chapters 2 and 3.

(1) System state. When all the external inputs are known, a set of information data that is known necessarily and its number is the fewest for determining the behaviors of a system are called the system state.

(2) State variables. Variables that can represent system state and change over time are called state variables and are usually expressed as $x_1(t)$, $x_2(t)$, $\ldots$, $x_n(t)$. Therefore, the system response value at any time t can be determined by the state variables and the excitations of the system at that time together. In addition, a state variable can also be said to be $\lambda_i(t)$.

Note that the selections of the state variables are usually not unique, but their number is unique, which means that the number of first-order differential equations forming the state equations is unique.

Although there are several different options for the state variables, they cannot be selected at will; all of them are required to satisfy independence and completeness. Independence means that none of the state variables can be represented by a linear combination of other state variables. Completeness means that it is impossible to find a state variable other than the determined variables.

(3) State vector. The column vector formed by the state variables $x_1(t)$, $x_2(t)$, $\ldots$, $x_n(t)$ is the state vector, and is represented by

$$\boldsymbol{x}(t) = \begin{bmatrix} x_1(t) \\ x_2(t) \\ \vdots \\ x_n(t) \end{bmatrix} = \begin{bmatrix} x_1(t) & x_2(t) & \cdots & x_n(t) \end{bmatrix}^T .$$

A state vector with n state variables is called the n-dimensional state vector.

(4) Representation of the state and the starting state. Values of the state variables at a certain time t_0 are the system state at t_0; its vector is expressed as

$$\mathbf{x}(t_0) = \begin{bmatrix} x_1(t_0) \\ x_2(t_0) \\ \vdots \\ x_n(t_0) \end{bmatrix} = \begin{bmatrix} x_1(t_0) & x_2(t_0) & \cdots & x_n(t_0) \end{bmatrix}^T .$$

Values of the state variables at the starting time 0_- are called the starting state; the vector is of the form

$$\mathbf{x}(0_-) = \begin{bmatrix} x_1(0_-) \\ x_2(0_-) \\ \vdots \\ x_n(0_-) \end{bmatrix} = \begin{bmatrix} x_1(0_-) & x_2(0_-) & \cdots & x_n(0_-) \end{bmatrix}^T .$$

It will subsequently be represented by $x(0)$ for convenience.

(5) State space. We can call the space that is used to place the state vectors as the state space. The projections of a state vector on n coordinate axes are the corresponding state variables.

(6) State equations. The state equations refer to a set of first-order differential equations, which can relate to the excitation vector $\mathbf{f}(t)$, the starting state vector $\mathbf{x}(0_-)$ and the state vector $\mathbf{x}(t)$ of an nth order system. They can reveal the interior change regular and motion state of the system.

(7) Output equations. The output equations refer to a set of algebraic equations to describe relations of the excitation vector $\mathbf{f}(t)$, the state vector $\mathbf{x}(t)$ and response vector $\mathbf{y}(t)$.

The advantages of using the state space method to analyze a system are the following.
(1) First-order differential equations are convenient for solving, especially for computer processing.
(2) Since the responses (outputs), the state variables and the excitations (inputs) are related by the algebraic equations (output equations), once the state variables are solved, each response can be determined by the linear combination of the state variables and the excitations.
(3) This can be extended easily in the analysis of time varying and nonlinear systems.

Note: The above conceptions are also suitable for discrete systems with tiny changes, such as t replaced by n, the differential equation replaced by difference equation, etc.

10.2 State equations of a system

The concept of the state space analysis method will be introduced by the following example.

Example 10.2-1. Write the state equations of a system shown in ▶ Figure 10.2.

Solution. Based on KCL and characteristics of inductor and capacitor, the equations are

$$Ri_L(t) + L\frac{di_L(t)}{dt} + u_C(t) = u_S(t), \tag{10.2-1}$$

$$C\frac{du_C(t)}{dt} = i_L(t). \tag{10.2-2}$$

If we are only concerned with the response–excitation relationship, substituting equation (10.2-2) into equation (10.2-1) yields

$$LC\frac{d^2u_C(t)}{dt^2} + RC\frac{du_C(t)}{dt} + u_C(t) = u_S(t). \tag{10.2-3}$$

Equation (10.2-3) is the familiar system model and is obtained by the external method. The response of the circuit to the given excitation can be obtained by solving this differential equation.

Now the problem is that we should know not only the relation between $u_C(t)$ and $u_S(t)$, but also the variation regular of $i_L(t)$ with $u_S(t)$, so equations (10.2-1) and (10.2-2) should be solved simultaneously as an equation set in the form

$$\begin{cases} \frac{d}{dt}i_L(t) = -\frac{R}{L}i_L(t) - \frac{1}{L}u_C(t) + \frac{1}{L}u_S(t) \\ \frac{d}{dt}u_C(t) = \frac{1}{C}i_L(t) \end{cases} \tag{10.2-4}$$

Equation (10.2-4) is a set of first-order simultaneous differential equations using $i_L(t)$ and $u_C(t)$ as the variables. As long as $u_S(t)$ is given and the starting states of $i_L(t)$ and $u_C(t)$ are known, the overall behaviors of the system can be totally understood. This method of analyzing the system is just the state space analysis method. So, equation (10.2-4) is the state equations of the system, and $i_L(t)$ and $u_C(t)$ are known as the state variables.

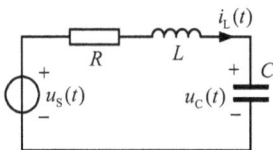

Fig. 10.2: E10.2-1.

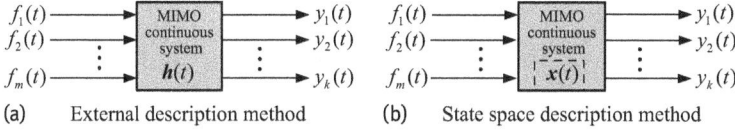

(a)   External description method   (b)   State space description method

Fig. 10.3: External and state space description method diagrams of a MIMO continuous system.

Usually, for the MIMO continuous system shown in ▶ Figure 10.3a, which has m inputs $f_1, f_2, \ldots, f_m$ and k outputs $y_1, y_2, \ldots, y_k$, when the starting conditions are slack, the equation of the system by external description is of the matrix form

$$
\begin{bmatrix} y_1(t) \\ y_2(t) \\ \vdots \\ y_n(t) \end{bmatrix} = \begin{bmatrix} h_{11}(t) & h_{12}(t) & \cdots & h_{1m}(t) \\ h_{21}(t) & h_{22}(t) & \cdots & h_{2m}(t) \\ \vdots & \vdots & & \vdots \\ h_{k1}(t) & h_{k2}(t) & \cdots & h_{km}(t) \end{bmatrix} * \begin{bmatrix} f_1(t) \\ f_2(t) \\ \vdots \\ f_n(t) \end{bmatrix}, \tag{10.2-5}
$$

or, simply,

$$
y(t) = h(t) * f(t). \tag{10.2-6}
$$

However, in the state space method it can be also described by a set of first-order differential equations (state equations) and a set of algebraic equations (output equations), that is,

$$
\begin{aligned}
\dot{x}_1 &= a_{11}x_1 + a_{12}x_2 + \cdots + a_{1n}x_n + b_{11}f_1 + b_{12}f_2 + \cdots + b_{1m}f_m \\
\dot{x}_2 &= a_{21}x_1 + a_{22}x_2 + \cdots + a_{2n}x_n + b_{21}f_1 + b_{22}f_2 + \cdots + b_{2m}f_m \\
&\vdots \\
\dot{x}_n &= a_{n1}x_1 + a_{n2}x_2 + \cdots + a_{nn}x_n + b_{n1}f_1 + b_{n2}f_2 + \cdots + b_{nm}f_m
\end{aligned} \tag{10.2-7}
$$

and

$$
\begin{aligned}
y_1 &= c_{11}x_1 + c_{12}x_2 + \cdots + c_{1n}x_n + d_{11}f_1 + d_{12}f_2 + \cdots + d_{1m}f_m \\
y_2 &= c_{21}x_1 + c_{22}x_2 + \cdots + c_{2n}x_n + d_{21}f_1 + d_{22}f_2 + \cdots + d_{2m}f_m \\
&\vdots \\
y_k &= c_{k1}x_1 + c_{k2}x_2 + \cdots + c_{kn}x_n + d_{k1}f_1 + d_{k2}f_2 + \cdots + d_{km}f_m
\end{aligned} \tag{10.2-8}
$$

where $\dot{x}_i = \frac{dx_i}{dt}$, x_i is the state variable, f_i is the input (excitation) and y_i is the response (output) of the system. equation (10.2-7) is the general form of the state equations, equation (10.2-8) is a general form of the output equations. If we write

$$x(t) = \begin{bmatrix} x_1(t) \\ x_2(t) \\ \vdots \\ x_n(t) \end{bmatrix}, \quad f(t) = \begin{bmatrix} f_1(t) \\ f_2(t) \\ \vdots \\ f_n(t) \end{bmatrix}, \quad y(t) = \begin{bmatrix} y_1(t) \\ y_2(t) \\ \vdots \\ y_n(t) \end{bmatrix}$$

$$A = \begin{bmatrix} a_{11} & a_{12} & \cdots & a_{1n} \\ a_{21} & a_{22} & \cdots & a_{2n} \\ \vdots & \vdots & & \vdots \\ a_{n1} & a_{n2} & \cdots & a_{nn} \end{bmatrix}, \quad B = \begin{bmatrix} b_{11} & b_{12} & \cdots & b_{1m} \\ b_{21} & b_{22} & \cdots & b_{2m} \\ \vdots & \vdots & & \vdots \\ b_{n1} & b_{n2} & \cdots & b_{nm} \end{bmatrix}$$

$$C = \begin{bmatrix} c_{11} & c_{12} & \cdots & c_{1n} \\ c_{21} & c_{22} & \cdots & c_{2n} \\ \vdots & \vdots & & \vdots \\ c_{k1} & c_{k2} & \cdots & c_{kn} \end{bmatrix}, \quad D = \begin{bmatrix} d_{11} & d_{12} & \cdots & d_{1m} \\ d_{21} & d_{22} & \cdots & d_{2m} \\ \vdots & \vdots & & \vdots \\ d_{k1} & d_{k2} & \cdots & d_{km} \end{bmatrix},$$

then the standard matrix forms of the state equations and the output equations can be expressed as

$$\begin{cases} \dot{x}(t) = Ax(t) + Bf(t) \\ y(t) = Cx(t) + Df(t) \end{cases},$$

or, simply,

$$\begin{cases} \dot{x} = Ax + Bf \\ y = Cx + Df \end{cases}, \qquad (10.2\text{-}9)$$

where A, B, C, D are constant matrices, respectively, in $n \times n$, $n \times m$, $k \times n$ and $k \times m$ orders.

Obviously, the state variables are the link between the excitations and responses. Equation (10.2-9) can also be called the standard form of the state model of a system and are roughly described by ▶ Figure 10.3b.

So far, we have known the basic concept of the state-space analysis method and the standard form of the state model, which includes the state equations and output equations. The next problem is how to establish the state model of a system.

10.3 Establishment of the state model

From the previous chapters, we know that a system can be described by using circuit diagrams, simulated block diagrams or flow graphs, mathematical models and system functions, etc. A new description method with the state model for a system is also introduced in this chapter, and we naturally ask: Can the state model be obtained based on the previous methods? The answer is "yes".

10.3.1 Establishment from circuit diagrams

As we know, an electric system (or electric network) generally mainly consists of resistors, inductors and capacitors. However, the voltage and the current of an inductor or a capacitor, which is the dynamic element, just satisfy the first-order differential relationship, so the two parameters of the two types of element are very easily related by the state equation and can precisely reflect the energy storage situation of the system. Hence, the voltage across a capacitor and the current through an inductor can frequently be chosen as the state variables.

The steps to directly list the state model from the circuit diagram are as follows:

(1) Select the voltages across all independent capacitors and currents through all independent inductors as the state variables.

(2) Use KCL to write the expressions of which each capacitor current $i_{C_i} = C_i \frac{du_{C_i}}{dt}$ relates to other state variables and inputs.

(3) Use KVL to write the expressions of which each inductor voltage $u_{L_i} = L_i \frac{di_{L_i}}{dt}$ relates to other state variables and inputs.

(4) If there are some non-state variables in KCL and KVL equations obtained from Step 2 and Step 3, these variables should be eliminated by using the KCL equations at appropriate nodes and the KVL equations in appropriate loops.

(5) Obtain the state equations of the circuit (system) by rearranging the relation expressions obtained from Steps 2 and 3 or Step 4 in the standard form.

(6) Obtain the output equations of the circuit (system) by relating the state variables and inputs and outputs based on the KCL and KVL equations.

The details of establishing the state model will be illustrated by the following two examples.

Example 10.3-1. Write the state model of the circuit shown in ▶ Figure 10.4, where $y_1(t)$ and $y_2(t)$ are the outputs, $f_1(t)$ is a current source and $f_2(t)$ is a voltage source.

Solution. Choose the inductor current and the capacitor voltage as the state variables $x_1(t)$ and $x_2(t)$. Based on KCL and KVL, we have

$$L\dot{x}_1 = y_1 - x_2 = R_1(f_1 - x_1) - x_2 = -R_1 x_1 - x_2 + R_1 f_1 ,$$

$$C\dot{x}_2 = x_1 + \frac{1}{R_2}(f_2 - x_2) = x_1 - \frac{1}{R_2}x_2 + \frac{1}{R_2}f_2 .$$

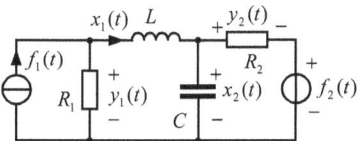

Fig. 10.4: System of E10.3-1.

These can be written in the state equation matrix form

$$
\begin{bmatrix} \dot{x}_1 \\ \dot{x}_2 \end{bmatrix} = \begin{bmatrix} -\frac{R_1}{L} & -\frac{1}{L} \\ \frac{1}{C} & -\frac{1}{R_2 C} \end{bmatrix} \begin{bmatrix} x_1 \\ x_2 \end{bmatrix} + \begin{bmatrix} \frac{R_1}{L} & 0 \\ 0 & \frac{1}{R_2 C} \end{bmatrix} \begin{bmatrix} f_1 \\ f_2 \end{bmatrix} .
$$

The output equations are

$$
y_1 = R_1(f_1 - x_1) = -R_1 x_1 + R_1 f_1 ,
$$
$$
y_2 = x_2 - f_2 .
$$

They can be written into the output equation matrix form

$$
\begin{bmatrix} y_1 \\ y_2 \end{bmatrix} = \begin{bmatrix} -R_1 & 0 \\ 0 & 1 \end{bmatrix} \begin{bmatrix} x_1 \\ x_2 \end{bmatrix} + \begin{bmatrix} R_1 & 0 \\ 0 & -1 \end{bmatrix} \begin{bmatrix} f_1 \\ f_2 \end{bmatrix} .
$$

Example 10.3-2. For the circuit shown in ▸ Figure 10.5, prove which can be considered as the state variables from the following options.
(1) $i_L(t), u_L(t)$ (2) $i_C(t), u_C(t)$ (3) $u_C(t), i_1(t)$ (4) $i_C(t), u_L(t)$

Solution. The selected variables must satisfy independence and completeness at the same time as state variables. Moreover, the response equations written from the selected variables should be linear algebraic equations.
(1) Since $u_L(t) = L\frac{di_L(t)}{dt}$, $u_L(t)$ and $i_L(t)$ are linearly independent (mutually independent), and according to the circuit diagram,

$$
u_C(t) = f(t) - u_L(t) - R_2 i_L(t)
$$
$$
i_2(t) = i_L(t)
$$
$$
u_2(t) = R_2 i_2(t) = R_2 i_L(t)
$$
$$
u_1(t) = u_2(t) + u_L(t) = R_2 i_L(t) + u_L(t)
$$
$$
i_1(t) = \frac{1}{R_1} u_1(t) = \frac{R_2}{R_1} i_L(t) + \frac{1}{R_1} u_L(t)
$$
$$
i_C(t) = i_1(t) + i_2(t) = \frac{1}{R_1} u_1(t) + i_L(t) = \frac{R_2}{R_1} i_L(t) + \frac{1}{R_1} u_L(t) + i_L(t)
$$

so, $i_L(t)$ and $u_L(t)$ can be state variables.

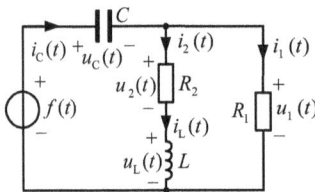

Fig. 10.5: System of E10.3-2.

(2) Since $i_C(t) = C\frac{du_C(t)}{dt}$, $u_C(t)$ and $i_C(t)$ are linearly independent, and according to the circuit diagram,

$$u_1(t) = f(t) - u_C(t)$$

$$i_1(t) = \frac{1}{R_1}u_1(t) = \frac{1}{R_1}f(t) - \frac{1}{R_1}u_C(t)$$

$$i_L(t) = i_C(t) - i_1(t) = i_C(t) - \frac{1}{R_1}f(t) - \frac{1}{R_1}u_C(t)$$

$$u_2(t) = R_2 i_L(t) = R_2 i_C(t) - \frac{R_2}{R_1}f(t) + \frac{R_2}{R_1}u_C(t)$$

$$i_2(t) = i_L(t) = i_C(t) - \frac{1}{R_1}f(t) - \frac{1}{R_1}u_C(t)$$

$$u_L(t) = f(t) - u_C(t) - u_2(t) = f(t) - u_C(t) - \left[R_2 i_C(t) - \frac{R_2}{R_1}f(t) + \frac{R_2}{R_1}u_C(t)\right]$$

So, $i_C(t)$ and $u_C(t)$ can also be state variables.

(3) Since $i_1(t) = \frac{1}{R_1}u_1(t) = \frac{1}{R_1}f(t) - \frac{1}{R_1}u_C(t)$, it is visible that $i_1(t)$ and $u_C(t)$ are linear correlated they cannot be state variables at the same time, but only one of them can be a state variable, and then a variable with independence and completeness will be re-selected as another state variable.

(4) Since

$$i_C(t) = i_1(t) + i_L(t) = \frac{1}{L}\int_{-\infty}^{t} u_L(\tau)d\tau + \frac{1}{R_1}u_1(t),$$

$$= \frac{1}{L}\int_{-\infty}^{t} u_L(\tau)d\tau + \frac{1}{R_1}[u_L(t) + u_2(t)]$$

$$= \frac{1}{L}\int_{-\infty}^{t} u_L(\tau)d\tau + \frac{1}{R_1}u_L(t) + \frac{1}{R_1}R_2 i_L(t)$$

$$= \frac{1}{L}\int_{-\infty}^{t} u_L(\tau)d\tau + \frac{1}{R_1}u_L(t) + \frac{R_2}{R_1}\frac{1}{L}\int_{-\infty}^{t} u_L(\tau)d\tau$$

It can be seen that $i_C(t)$ and $u_L(t)$ are linear independent, so they can be state variables. This example states that **the choice of state variables is not unique.**

10.3.2 Establishment method from emulation diagrams

We know that an important part in the emulation diagram is the integrator. Therefore, we should deal with it before we establish the state model by emulation diagrams. Assuming that an integrator output is x_o and the input is x_i,

$$\dot{x}_o = x_i \qquad (10.3\text{-}1)$$

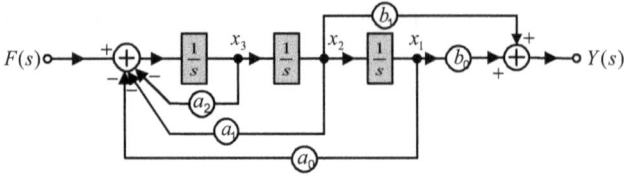

Fig. 10.6: E10.3-3.

Obviously, the integrator output can be used as a state variable, which is the key to establishing the state model. Establishing the state model from emulation diagrams (block or flow diagrams) is easier and more intuitive than the method using circuit diagrams. The general procedure is as follows.

(1) Select each integrator output (or differentiator input) as a state variable.
(2) List the state equations and the output equations based on each adder.

Example 10.3-3. Establish the state equations and the output equations of the third-order system shown in ▶ Figure 10.6.

Solution. Select outputs of three integrators as state variables such as x_1, x_2, x_3. The state equations around the first adder are the following.

$$\dot{x}_1 = x_2$$
$$\dot{x}_2 = x_3$$
$$\dot{x}_3 = -a_0 x_1 - a_1 x_2 - a_2 x_3 + f$$

Around the second adder, the output equation is

$$y = b_0 x_1 + b_1 x_2 .$$

All the above equations can be expressed in matrix form

$$\begin{bmatrix} \dot{x}_1 \\ \dot{x}_2 \\ \dot{x}_3 \end{bmatrix} = \begin{bmatrix} 0 & 1 & 0 \\ 0 & 0 & 1 \\ -a_0 & -a_1 & -a_2 \end{bmatrix} \begin{bmatrix} x_1 \\ x_2 \\ x_3 \end{bmatrix} + \begin{bmatrix} 0 \\ 0 \\ 1 \end{bmatrix} [f] ,$$

$$[y] = \begin{bmatrix} b_0 & b_1 & 0 \end{bmatrix} \begin{bmatrix} x_1 \\ x_2 \\ x_3 \end{bmatrix} .$$

Example 10.3-4. ▶ Figure 10.7 is a system flow graph. Establish its state model.

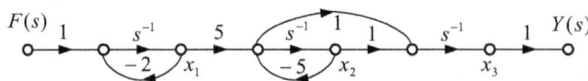

Fig. 10.7: E10.3-4.

Solution. Selecting three integrator outputs as the state variables, we have

$$\dot{x}_1 = -2x_1 + f$$
$$\dot{x}_2 = 5x_1 - 5x_2$$
$$\dot{x}_3 = \dot{x}_2 + x_2 = 5x_1 - 5x_2 + x_2 = 5x_1 - 4x_2$$

Writing them matrix form

$$\begin{bmatrix} \dot{x}_1 \\ \dot{x}_2 \\ \dot{x}_3 \end{bmatrix} = \begin{bmatrix} -2 & 0 & 0 \\ 5 & -5 & 0 \\ 5 & -4 & 0 \end{bmatrix} \begin{bmatrix} x_1 \\ x_2 \\ x_3 \end{bmatrix} + \begin{bmatrix} 1 \\ 0 \\ 0 \end{bmatrix} [f].$$

The output equation is

$$y = x_3.$$

That is,

$$[y] = \begin{bmatrix} 0 & 0 & 1 \end{bmatrix} \begin{bmatrix} x_1 \\ x_2 \\ x_3 \end{bmatrix}.$$

It can be seen from the two examples that regardless of whether the system is represented by a block diagram or a flow graph, the steps or methods of establishing the state equations and the output equations are the same, because the flow graph is actually a simplified form of the block diagram, and there is no essential difference between them.

10.3.3 Establishment method from mathematical models

The mathematical models of the continuous system we are familiar with are mainly the linear differential equation and the system function $H(s)$. Then, how can we establish the state model from the mathematical models? Let us look at the following example.

Example 10.3-5. $H(s) = \frac{b_1 s + b_0}{s^3 + a_2 s^2 + a_1 s + a_0}$ is given as the system function. Establish the state equations and output equations of the system.

Solution. From the system function, the two direct forms, the parallel and serial flow graphs are plotted in ▶ Figure 10.8.
From the direct form 1, we can obtain

$$\dot{x}_1 = x_2$$
$$\dot{x}_2 = x_3$$
$$\dot{x}_3 = -a_0 x_1 - a_1 x_2 - a_2 x_3 + f$$
$$y = b_0 x_1 + b_1 x_2$$

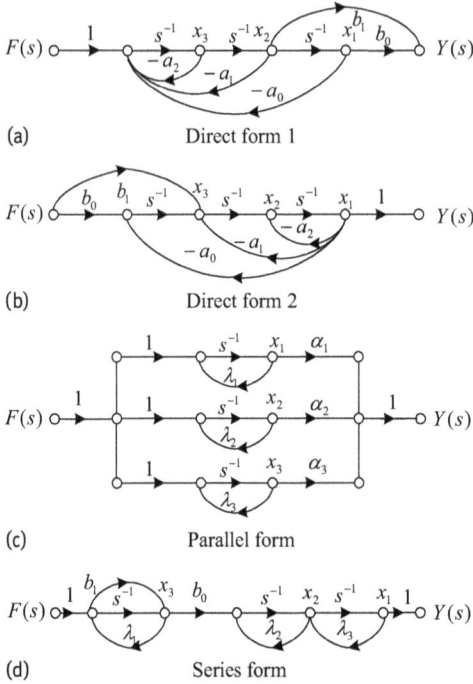

(a) Direct form 1

(b) Direct form 2

(c) Parallel form

(d) Series form

Fig. 10.8: E10.3-5.

Writing them in the matrix form

$$
\begin{bmatrix} \dot{x}_1 \\ \dot{x}_2 \\ \dot{x}_3 \end{bmatrix} = \begin{bmatrix} 0 & 1 & 0 \\ 0 & 0 & 1 \\ -a_0 & -a_1 & -a_2 \end{bmatrix} \begin{bmatrix} x_1 \\ x_2 \\ x_3 \end{bmatrix} + \begin{bmatrix} 0 \\ 0 \\ 1 \end{bmatrix} [f],
$$

$$
[y] = \begin{bmatrix} b_0 & b_1 & 0 \end{bmatrix} \begin{bmatrix} x_1 \\ x_2 \\ x_3 \end{bmatrix}.
$$

This is just the system in Example 10.3-3.

From the direct form 2, we can obtain

$$
\dot{x}_1 = -a_2 x_1 + x_2
$$
$$
\dot{x}_2 = -a_1 x_1 + x_3 + b_1 f
$$
$$
\dot{x}_3 = -a_0 x_1 + b_0 f
$$
$$
y = x_1 .
$$

Their matrix forms are

$$
\begin{bmatrix} \dot{x}_1 \\ \dot{x}_2 \\ \dot{x}_3 \end{bmatrix} = \begin{bmatrix} -a_2 & 1 & 0 \\ -a_1 & 0 & 1 \\ -a_0 & 0 & 0 \end{bmatrix} \begin{bmatrix} x_1 \\ x_2 \\ x_3 \end{bmatrix} + \begin{bmatrix} 0 \\ b_1 \\ b_0 \end{bmatrix} \begin{bmatrix} f \end{bmatrix},
$$

$$
\begin{bmatrix} y \end{bmatrix} = \begin{bmatrix} 1 & 0 & 0 \end{bmatrix} \begin{bmatrix} x_1 \\ x_2 \\ x_3 \end{bmatrix}.
$$

For the parallel form, we have $H(s) = \frac{\alpha_1}{s-\lambda_1} + \frac{\alpha_2}{s-\lambda_2} + \frac{\alpha_3}{s-\lambda_3}$, where $\lambda_1, \lambda_2, \lambda_3$ are dissimilar simple roots. Then,

$$
\dot{x}_1 = \lambda_1 x_1 + f
$$
$$
\dot{x}_2 = \lambda_2 x_2 + f
$$
$$
\dot{x}_3 = \lambda_3 x_3 + f
$$
$$
y = \alpha_1 x_1 + \alpha_2 x_2 + \alpha_3 x_3
$$

Their matrix forms are

$$
\begin{bmatrix} \dot{x}_1 \\ \dot{x}_2 \\ \dot{x}_3 \end{bmatrix} = \begin{bmatrix} \lambda_1 & 0 & 0 \\ 0 & \lambda_2 & 0 \\ 0 & 0 & \lambda_3 \end{bmatrix} \begin{bmatrix} x_1 \\ x_2 \\ x_3 \end{bmatrix} + \begin{bmatrix} 1 \\ 1 \\ 1 \end{bmatrix} \begin{bmatrix} f \end{bmatrix}
$$

$$
\begin{bmatrix} y \end{bmatrix} = \begin{bmatrix} \alpha_1 & \alpha_2 & \alpha_2 \end{bmatrix} \begin{bmatrix} x_1 \\ x_2 \\ x_3 \end{bmatrix}
$$

The matrix A in the parallel form is a diagonal matrix. The elements on the diagonal line are the characteristic roots of the system, so state variables in parallel form are also called diagonal variables.

For the serial form, we have

$$
H(s) = \frac{b_1 s + b_0}{s - \lambda_1} \cdot \frac{1}{s - \lambda_2} \cdot \frac{1}{s - \lambda_3}.
$$

We can obtain

$$
\begin{bmatrix} \dot{x}_1 \\ \dot{x}_2 \\ \dot{x}_3 \end{bmatrix} = \begin{bmatrix} \lambda_3 & 1 & 0 \\ 0 & \lambda_2 & b_0 + b_1 \lambda_1 \\ 0 & 0 & \lambda_1 \end{bmatrix} \begin{bmatrix} x_1 \\ x_2 \\ x_3 \end{bmatrix} + \begin{bmatrix} 0 \\ b_1 \\ 1 \end{bmatrix} \begin{bmatrix} f \end{bmatrix},
$$

$$
\begin{bmatrix} y \end{bmatrix} = x_1.
$$

The matrix A in serial form is a triangular matrix, and the diagonal elements are also the characteristic roots of the system.

The example gives the following points:

(1) A system has different state models in form, which means that both forms of the state equations and the output equations are not unique.

(2) Based on the following system function form

$$H(s) = \frac{b_m s^m + b_{m-1} s^{m-1} + \cdots + b_1 s + b_0}{s^n + a_{n-1} s^{n-1} + \cdots + a_1 s + a_0}, \tag{10.3-2}$$

three state model forms can be listed.

The first form:

$$
\begin{bmatrix} \dot{x}_1 \\ \dot{x}_2 \\ \vdots \\ \dot{x}_{n-1} \\ \dot{x}_n \end{bmatrix}
=
\begin{bmatrix}
0 & 1 & 0 & \cdots & 0 & 0 \\
0 & 0 & 1 & 0 & \cdots & 0 \\
\vdots & \vdots & \vdots & \vdots & \vdots & \vdots \\
0 & 0 & 0 & 0 & \cdots & 1 \\
-a_0 & -a_1 & -a_2 & \cdots & -a_{n-2} & -a_{n-1}
\end{bmatrix}
\begin{bmatrix} x_1 \\ x_2 \\ \vdots \\ x_{n-1} \\ x_n \end{bmatrix}
+
\begin{bmatrix} 0 \\ 0 \\ 0 \\ \vdots \\ 1 \end{bmatrix} [f] \tag{10.3-3}
$$

$$
[y] = \begin{bmatrix} b_0 & b_1 & b_2 & \cdots & b_m & 0 & \cdots & 0 \end{bmatrix}
\begin{bmatrix} x_1 \\ x_2 \\ \vdots \\ x_{n-1} \\ x_n \end{bmatrix} \tag{10.3-4}
$$

The second form is

$$
\begin{bmatrix} \dot{x}_1 \\ \dot{x}_2 \\ \vdots \\ \dot{x}_{n-1} \\ \dot{x}_n \end{bmatrix}
=
\begin{bmatrix}
-a_{n-1} & 1 & 0 & \cdots & 0 & 0 \\
-a_{n-2} & 0 & 1 & 0 & \cdots & 0 \\
\vdots & \vdots & \vdots & \vdots & \vdots & \vdots \\
-a_1 & 0 & 0 & 0 & \cdots & 1 \\
-a_0 & 0 & 0 & \cdots & 0 & 0
\end{bmatrix}
\begin{bmatrix} x_1 \\ x_2 \\ \vdots \\ x_{n-1} \\ x_n \end{bmatrix}
+
\begin{bmatrix} 0 \\ \vdots \\ 0 \\ b_m \\ \vdots \\ b_1 \\ b_0 \end{bmatrix} [f] \tag{10.3-5}
$$

$$
[y] = \begin{bmatrix} 1 & 0 & 0 & \cdots & 0 \end{bmatrix}
\begin{bmatrix} x_1 \\ x_2 \\ \vdots \\ x_{n-1} \\ x_n \end{bmatrix} \tag{10.3-6}
$$

The third form is

$$
\begin{bmatrix} \dot{x}_1 \\ \dot{x}_2 \\ \vdots \\ \dot{x}_{n-1} \\ \dot{x}_n \end{bmatrix} = \begin{bmatrix} \lambda_1 & 0 & 0 & \cdots & 0 \\ 0 & \lambda_2 & 0 & \cdots & 0 \\ 0 & 0 & \ddots & \cdots & \vdots \\ \vdots & \vdots & \vdots & \ddots & 0 \\ 0 & 0 & \cdots & 0 & \lambda_n \end{bmatrix} \begin{bmatrix} x_1 \\ x_2 \\ \vdots \\ x_{n-1} \\ x_n \end{bmatrix} + \begin{bmatrix} 1 \\ 1 \\ 1 \\ \vdots \\ 1 \end{bmatrix} [f] \tag{10.3-7}
$$

$$
[y] = \begin{bmatrix} K_1 & K_2 & K_3 & \cdots & K_n \end{bmatrix} \begin{bmatrix} x_1 \\ x_2 \\ \vdots \\ x_{n-1} \\ x_n \end{bmatrix}, \tag{10.3-8}
$$

where $K_1, K_2, \ldots, K_n$ are, respectively, the coefficients of the numerator in each partial fraction expression of $H(s)$; then

$$
H(s) = \frac{K_1}{s - \lambda_1} + \frac{K_2}{s - \lambda_2} + \cdots + \frac{K_n}{s - \lambda_n}. \tag{10.3-9}
$$

Note: In the third form, $H(s)$ should have n different simple roots.

It should be explained that, in this section, mainly the methods to establish the state models for a continuous system are discussed, but the establishment and methods for the state models of a discrete system are similar to them. For example, the steps to establish the state model from the discrete system diagram are as follows.
(1) Select each delayer output as the state variable.
(2) List the state or output equations around each adder.

The standard form of the discrete state model is of the form

$$
\begin{cases} x[n + 1] = Ax[n] + Bf[n] \\ y[n] = Cx[n] + Df[n] \end{cases} \tag{10.3-10}
$$

Example 10.3-6. The mathematical model for a discrete system is given by

$$
y[n + 3] + 8y[n + 2] + 16y[n + 1] + 10y[n] = 6f[n + 2] + 12f[n + 1] + 18f[n] .
$$

Find its state model.

Solution. The transfer operator $H(E) = \frac{6E^2 + 12E + 18}{E^3 + 8E^2 + 16E + 10}$ is obtained from the difference equation; the flow graph is shown in ▸ Figure 10.9 by Mason's formula.
The state equations are

$$
x_1[n + 1] = x_2[n]
$$
$$
x_2[n + 1] = x_3[n]
$$
$$
x_3[n + 1] = -10x_1[n] - 16x_2[n] - 8x_3[n] + f[n]
$$

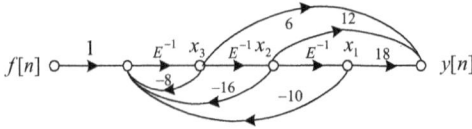

Fig. 10.9: E10.3-6.

The output equation is

$$y[n] = 18x_1[n] + 12x_2[n] + 6x_3[n]$$

Writing them in the matrix form,

$$\begin{bmatrix} x_1[n+1] \\ x_2[n+1] \\ x_3[n+1] \end{bmatrix} = \begin{bmatrix} 0 & 1 & 0 \\ 0 & 0 & 1 \\ -10 & -16 & -8 \end{bmatrix} \begin{bmatrix} x_1[n] \\ x_2[n] \\ x_3[n] \end{bmatrix} + \begin{bmatrix} 0 \\ 0 \\ 1 \end{bmatrix} f[n],$$

$$y[n] = \begin{bmatrix} 18 & 12 & 6 \end{bmatrix} \begin{bmatrix} x_1[n] \\ x_2[n] \\ x_3[n] \end{bmatrix}.$$

The details about the state space analysis for discrete systems can be found in related books.

10.4 Solutions of the state model

Because the state equations are a set of simultaneous first-order differential equations, obviously, they can be solved in both the time and s domains. The solution methods in the s domain are easier than those in the time domain, so they will be introduced first here.

10.4.1 Solutions in the s domain

The essential point of solving methods for the state equations in the s domain is the same as with the methods for the differential equation in the s domain. Let us study the following state equations:

$$\dot{x}_1 = a_{11}x_1 + a_{12}x_2 + \cdots + a_{1n}x_n + b_{11}f_1 + b_{12}f_2 + \cdots + b_{1m}f_m$$
$$\dot{x}_2 = a_{21}x_1 + a_{22}x_2 + \cdots + a_{2n}x_n + b_{21}f_1 + b_{22}f_2 + \cdots + b_{2m}f_m$$
$$\vdots$$
$$\dot{x}_n = a_{n1}x_1 + a_{n2}x_2 + \cdots + a_{nn}x_n + b_{n1}f_1 + b_{n2}f_2 + \cdots + b_{nm}f_m$$

The kth equation among them can be written as

$$\dot{x}_k = a_{k1}x_1 + a_{k2}x_2 + \cdots + a_{kn}x_n + b_{k1}f_1 + b_{k2}f_2 + \cdots + b_{km}f_m . \qquad (10.4\text{-}1)$$

With the Laplace transform on both sides of equation (10.4-1), we have

$$sX_k(s) - x_k(0) = a_{k1}X_1(s) + a_{k2}X_2(s) + \cdots + a_{kn}X_n(s) + b_{k1}F_1(s) + b_{k2}F_2(s) + \cdots + b_{km}F_m(s).$$

Then the Laplace transforms for all the state equations can be obtained according to the process,

$$s \underbrace{\begin{bmatrix} X_1(s) \\ X_2(s) \\ \vdots \\ X_n(s) \end{bmatrix}}_{X(s)} - \underbrace{\begin{bmatrix} x_1(0) \\ x_2(0) \\ \vdots \\ x_n(0) \end{bmatrix}}_{x(0)}$$

$$= \underbrace{\begin{bmatrix} a_{11} & a_{12} & \cdots & a_{1n} \\ a_{21} & a_{22} & \cdots & a_{2n} \\ \vdots & \vdots & \vdots & \vdots \\ a_{n1} & a_{n2} & \cdots & a_{nn} \end{bmatrix}}_{A} \underbrace{\begin{bmatrix} X_1(s) \\ X_2(s) \\ \vdots \\ X_n(s) \end{bmatrix}}_{X(s)} + \underbrace{\begin{bmatrix} b_{11} & b_{12} & \cdots & b_{1m} \\ b_{21} & b_{22} & \cdots & b_{2m} \\ \vdots & \vdots & \vdots & \vdots \\ b_{n1} & b_{n2} & \cdots & b_{nm} \end{bmatrix}}_{B} \underbrace{\begin{bmatrix} F_1(s) \\ F_2(s) \\ \vdots \\ F_m(s) \end{bmatrix}}_{F(s)} \qquad (10.4\text{-}2)$$

Writing equation (10.4-2) in the vector form,

$$sX(s) - x(0) = AX(s) + BF(s) . \qquad (10.4\text{-}3)$$

Rearranging terms yields

$$(sI - A)X(s) = x(0) + BF(s) ,$$

where I is a unit matrix. The state vector can be solved from the above equation, yielding

$$X(s) = (sI - A)^{-1}[x(0) + BF(s)] = \Phi(s)[x(0) + BF(s)] ,$$

namely,

$$X(s) = \Phi(s)x(0) + \Phi(s)BF(s) . \qquad (10.4\text{-}4)$$

In equation (10.4-4), we define $\Phi(s) = (sI - A)^{-1}$ as the decomposition matrix or the preliminary solution matrix.

With the inverse Laplace transform on both sides of equation (10.4-4), we can obtain the state vector solution in the time domain

$$x(t) = \underbrace{\mathcal{L}^{-1}[\Phi(s)]x(0)}_{\substack{\text{zero input} \\ \text{component}}} + \underbrace{\mathcal{L}^{-1}[\Phi(s)BF(s)]}_{\substack{\text{zero state} \\ \text{component}}} . \qquad (10.4\text{-}5)$$

Obviously, the first term in equation (10.4-5) is only related to the starting state $x(0)$, and when $x(0) = 0$, this component is zero, so it can be seen as the zero-input component. The second term is a function of the input $F(s)$, so it can be seen as the zero-state component.

Now, we will introduce the solving process for the output equations. From the related content in Section 10.2, the output equations are given by

$$y(t) = Cx(t) + Df(t) \, .$$

The expression in the s domain is

$$Y(s) = CX(s) + DF(s) \, .$$

Substituting equation (10.4-4) into the equation, we have

$$Y(s) = \underbrace{C\Phi(s)x(0)}_{\substack{\text{zero input} \\ \text{component}}} + \underbrace{[C\Phi(s)B + D]F(s)}_{\substack{\text{zero state} \\ \text{component}}} \, . \tag{10.4-6}$$

The zero-state response can be taken from equation (10.4-6) as

$$Y_f(s) = [C\Phi(s)B + D]F(s) \, . \tag{10.4-7}$$

The zero-state response and the excitation satisfy the following relation:

$$Y_f(s) = H(s)F(s) \, . \tag{10.4-8}$$

Comparing equations (10.4-7) and (10.4-8), we obtain the system function in matrix form,

$$H(s) = C\Phi(s)B + D \, . \tag{10.4-9}$$

The $H(s)$ is a $k \times m$-order matrix, with outputs k and inputs m, and all its elements are represented as $H_{ij}(s)$. The $H_{ij}(s)$ is the transfer function (system function) which joins the ith output $Y_i(s)$ and the jth input $F_j(s)$, such as $H_{ij}(s) = \frac{Y_i(s)}{F_j(s)}$.

From equations (10.4-6) and (10.4-9), the solution of system response in the time domain can be written as

$$y(t) = \underbrace{\mathcal{L}^{-1}[C\Phi(s)x(0)]}_{\substack{\text{zero input} \\ \text{component}}} + \underbrace{\mathcal{L}^{-1}[H(s)F(s)]}_{\substack{\text{zero state} \\ \text{component}}} \, . \tag{10.4-10}$$

Example 10.4-1. The state equation of a system is $\dot{x}(t) = Ax(t) + Bf(t)$, where

$$A = \begin{bmatrix} -12 & \frac{2}{3} \\ -36 & -1 \end{bmatrix}, \quad B = \begin{bmatrix} \frac{1}{3} \\ 1 \end{bmatrix},$$

and the starting conditions are $x_1(0) = 2$, $x_2(0) = 1$, and $f(t) = \varepsilon(t)$. Find the state vector $x(t)$.

Solution. Because

$$x(0) = \begin{bmatrix} 2 \\ 1 \end{bmatrix}, \quad F(s) = \frac{1}{s},$$

and

$$sI - A = s\begin{bmatrix} 1 & 0 \\ 0 & 1 \end{bmatrix} - \begin{bmatrix} -12 & \frac{2}{3} \\ -36 & -1 \end{bmatrix} = \begin{bmatrix} s+12 & -\frac{2}{3} \\ 36 & s+1 \end{bmatrix},$$

$$\Phi(s) = (sI - A)^{-1} = \begin{bmatrix} \frac{s+1}{(s+4)(s+9)} & \frac{2/3}{(s+4)(s+9)} \\ \frac{-36}{(s+4)(s+9)} & \frac{s+12}{(s+4)(s+9)} \end{bmatrix},$$

so,

$$X(s) = \Phi(s)x(0) + \Phi(s)BF(s)$$

$$= \begin{bmatrix} \frac{s+1}{(s+4)(s+9)} & \frac{2/3}{(s+4)(s+9)} \\ \frac{-36}{(s+4)(s+9)} & \frac{s+12}{(s+4)(s+9)} \end{bmatrix}\begin{bmatrix} 2 \\ 1 \end{bmatrix} + \begin{bmatrix} \frac{s+1}{(s+4)(s+9)} & \frac{2/3}{(s+4)(s+9)} \\ \frac{-36}{(s+4)(s+9)} & \frac{s+12}{(s+4)(s+9)} \end{bmatrix}\begin{bmatrix} \frac{1}{3} \\ 1 \end{bmatrix}\frac{1}{s}$$

$$= \begin{bmatrix} \frac{s+1}{(s+4)(s+9)} & \frac{2/3}{(s+4)(s+9)} \\ \frac{-36}{(s+4)(s+9)} & \frac{s+12}{(s+4)(s+9)} \end{bmatrix}\begin{bmatrix} \frac{6s+1}{3s} \\ \frac{s+1}{s} \end{bmatrix} = \begin{bmatrix} \frac{2s^2+3s+1}{s(s+4)(s+9)} \\ \frac{s-59}{(s+4)(s+9)} \end{bmatrix}$$

$$= \begin{bmatrix} \frac{1/36}{s} - \frac{21/20}{s+4} + \frac{136/45}{s+9} \\ \frac{-63/5}{s+4} + \frac{68/5}{s+9} \end{bmatrix}.$$

Applying the inverse Laplace transform to the equation,

$$\begin{bmatrix} x_1(t) \\ x_2(t) \end{bmatrix} = \begin{bmatrix} \left(\frac{1}{36} - \frac{21}{20}e^{-4t} + \frac{136}{45}e^{-9t}\right)\varepsilon(t) \\ \left(-\frac{63}{5}e^{-4t} + \frac{68}{5}e^{-9t}\right)\varepsilon(t) \end{bmatrix}.$$

Example 10.4-2. If the state and output equations of a system, respectively, are

$$\begin{bmatrix} \dot{x}_1 \\ \dot{x}_2 \end{bmatrix} = \begin{bmatrix} 0 & 1 \\ -2 & -3 \end{bmatrix}\begin{bmatrix} x_1 \\ x_2 \end{bmatrix} + \begin{bmatrix} 1 & 0 \\ 1 & 1 \end{bmatrix}\begin{bmatrix} f_1 \\ f_2 \end{bmatrix},$$

$$\begin{bmatrix} y_1 \\ y_2 \\ y_3 \end{bmatrix} = \begin{bmatrix} 1 & 0 \\ 1 & 1 \\ 0 & 2 \end{bmatrix}\begin{bmatrix} x_1 \\ x_2 \end{bmatrix} + \begin{bmatrix} 0 & 0 \\ 1 & 0 \\ 0 & 1 \end{bmatrix}\begin{bmatrix} f_1 \\ f_2 \end{bmatrix}.$$

Solve the system function $H(s)$ and the transfer function $H_{32}(s)$ linking output $y_3(t)$ and input $f_2(t)$.

Solution. From the given questions,

$$A = \begin{bmatrix} 0 & 1 \\ -2 & -3 \end{bmatrix}, \quad B = \begin{bmatrix} 1 & 0 \\ 1 & 1 \end{bmatrix}, \quad C = \begin{bmatrix} 1 & 0 \\ 1 & 1 \\ 0 & 2 \end{bmatrix}, \quad D = \begin{bmatrix} 0 & 0 \\ 1 & 0 \\ 0 & 1 \end{bmatrix}.$$

Then,

$$\Phi(s) = (sI - A)^{-1} = \begin{bmatrix} \frac{s+3}{(s+1)(s+2)} & \frac{1}{(s+1)(s+2)} \\ \frac{-2}{(s+1)(s+2)} & \frac{s}{(s+1)(s+2)} \end{bmatrix}.$$

By the formula $H(s) = C\Phi(s)B + D$ we obtain

$$H(s) = \begin{bmatrix} 1 & 0 \\ 1 & 1 \\ 0 & 2 \end{bmatrix} \begin{bmatrix} \frac{s+3}{(s+1)(s+2)} & \frac{1}{(s+1)(s+2)} \\ \frac{-2}{(s+1)(s+2)} & \frac{s}{(s+1)(s+2)} \end{bmatrix} \begin{bmatrix} 1 & 0 \\ 1 & 1 \end{bmatrix} + \begin{bmatrix} 0 & 0 \\ 1 & 0 \\ 0 & 1 \end{bmatrix}$$

$$= \begin{bmatrix} \frac{s+4}{(s+1)(s+2)} & \frac{1}{(s+1)(s+2)} \\ \frac{s+4}{s+2} & \frac{1}{s+2} \\ \frac{2(s-2)}{(s+1)(s+2)} & \frac{s^2+5s+2}{(s+1)(s+2)} \end{bmatrix}.$$

Because the element in row i and column j in the transfer function matrix is the transfer function of contacting the output i and the input j, from the equation, the transfer function of contacting the output $y_3(t)$ and the input $f_2(t)$ is

$$H_{32}(s) = \frac{s^2 + 5s + 2}{(s + 1)(s + 2)}.$$

This example states that because $\Phi(s) = (sI - A)^{-1}$, the denominators of elements in $\Phi(s)$ are all $|sI - A|$, and these denominators are also the denominators of elements in $H(s)$. Thus, the poles of the transfer function of the system are the zeros of the polynomial $|sI - A|$, which are also the natural frequencies of the system. So, the characteristic equation of the system is

$$|sI - A| = 0 \tag{10.4-11}$$

We already know the relations between the characteristic roots and the zero-input response, so if n distinct simple roots are worked out by equation (10.4-11), such as $\lambda_1, \lambda_2, \ldots, \lambda_n$, the zero-input response of the system is of the form

$$y_x(t) = c_1 e^{\lambda_1 t} + c_2 e^{\lambda_2 t} + \cdots + c_n e^{\lambda_n t} \tag{10.4-12}$$

Example 10.4-3. The state equations and the output equations of a system are

$$\begin{bmatrix} \dot{x}_1 \\ \dot{x}_2 \end{bmatrix} = \begin{bmatrix} 2 & 3 \\ 0 & -1 \end{bmatrix} \begin{bmatrix} x_1 \\ x_2 \end{bmatrix} + \begin{bmatrix} 0 & 1 \\ 1 & 0 \end{bmatrix} \begin{bmatrix} f_1 \\ f_2 \end{bmatrix}, \quad \begin{bmatrix} y_1 \\ y_2 \end{bmatrix} = \begin{bmatrix} 1 & 1 \\ 0 & -1 \end{bmatrix} \begin{bmatrix} x_1 \\ x_2 \end{bmatrix} + \begin{bmatrix} 1 & 0 \\ 1 & 0 \end{bmatrix} \begin{bmatrix} f_1 \\ f_2 \end{bmatrix}$$

The starting states and inputs are

$$\begin{bmatrix} x_1(0_-) \\ x_2(0_-) \end{bmatrix} = \begin{bmatrix} 2 \\ -1 \end{bmatrix}, \quad \begin{bmatrix} f_1 \\ f_2 \end{bmatrix} = \begin{bmatrix} \varepsilon(t) \\ \delta(t) \end{bmatrix}.$$

Find the state variables and outputs of the system.

Solution.

$$\Phi(s) = (sI - A)^{-1} = \begin{bmatrix} \frac{1}{s-2} & \frac{3}{(s+1)(s-2)} \\ 0 & \frac{1}{s+1} \end{bmatrix}$$

By the formula $X(s) = \boldsymbol{\Phi}(s)x(0) + \boldsymbol{\Phi}(s)BF(s)$,

$$\begin{bmatrix} X_1(s) \\ X_2(s) \end{bmatrix} = \begin{bmatrix} \frac{1}{s-2} & \frac{3}{(s+1)(s-2)} \\ 0 & \frac{1}{s+1} \end{bmatrix} \begin{bmatrix} 2 \\ -1 \end{bmatrix} + \begin{bmatrix} \frac{1}{s-2} & \frac{3}{(s+1)(s-2)} \\ 0 & \frac{1}{s+1} \end{bmatrix} \begin{bmatrix} 0 & 1 \\ 1 & 0 \end{bmatrix} \begin{bmatrix} \frac{1}{s} \\ 1 \end{bmatrix}$$

$$= \begin{bmatrix} \frac{1}{s-2} + \frac{1}{s+1} \\ \frac{-1}{s+1} \end{bmatrix} + \begin{bmatrix} \frac{3}{2}\frac{1}{s-2} + \frac{1}{s+1} - \frac{3}{2s} \\ \frac{1}{s} - \frac{1}{s+1} \end{bmatrix}$$

By the formula $Y(s) = CX(s) + DF(s)$,

$$\begin{bmatrix} Y_1(s) \\ Y_2(s) \end{bmatrix} = \begin{bmatrix} 1 & 1 \\ 0 & -1 \end{bmatrix} \left\{ \begin{bmatrix} \frac{1}{s-2} + \frac{1}{s+1} \\ \frac{-1}{s+1} \end{bmatrix} + \begin{bmatrix} \frac{3}{2}\frac{1}{s-2} + \frac{1}{s+1} - \frac{3}{2s} \\ \frac{1}{s} - \frac{1}{s+1} \end{bmatrix} \right\} + \begin{bmatrix} 1 & 0 \\ 1 & 0 \end{bmatrix} \begin{bmatrix} \frac{1}{s} \\ 1 \end{bmatrix}$$

$$= \begin{bmatrix} \frac{1}{s-2} \\ \frac{1}{s+1} \end{bmatrix} + \begin{bmatrix} \frac{3}{2}\frac{1}{s-2} + \frac{1}{2s} \\ \frac{1}{s+1} \end{bmatrix}$$

Applying the inverse Laplace transform to the two equations above yields

$$\begin{bmatrix} x_1(t) \\ x_2(t) \end{bmatrix} = \begin{bmatrix} e^{2t} + e^{-t} \\ -e^{-t} \end{bmatrix} + \begin{bmatrix} \frac{3}{2}e^{2t} + e^{-t} - \frac{3}{2} \\ 1 - e^{-t} \end{bmatrix} = \begin{bmatrix} \frac{5}{2}e^{2t} + 2e^{-t} - \frac{3}{2} \\ 1 - 2e^{-t} \end{bmatrix} \quad (t > 0)$$

$$\begin{bmatrix} y_1(t) \\ y_2(t) \end{bmatrix} = \underbrace{\begin{bmatrix} e^{2t} \\ e^{-t} \end{bmatrix}}_{\substack{\text{zero input} \\ \text{component}}} + \underbrace{\begin{bmatrix} \frac{3}{2}e^{2t} + \frac{1}{2} \\ e^{-t} \end{bmatrix}}_{\substack{\text{zero state} \\ \text{component}}} = \underbrace{\begin{bmatrix} \frac{5}{2}e^{2t} + \frac{1}{2} \\ 2e^{-t} \end{bmatrix}}_{\substack{\text{complete} \\ \text{response}}} \quad (t > 0)$$

In summary, the general steps of state space system analysis in the s domain can be listed as the following.

(1) The solution for the state equations.
The state equation and the starting state are, respectively, given as $\dot{x}(t) = Ax(t) + Bf(t)$ and $x(0)$. Find $x(t)$.
Step 1: $sI - A = ?$
Step 2: $\boldsymbol{\Phi}(s) = (sI - A)^{-1} = ?$
Step 3: $F(s) = ?$
Step 4: $X(s) = \boldsymbol{\Phi}(s)x(0) + \boldsymbol{\Phi}(s)BF(s) = ?$
Step 5: $x(t) = \mathcal{L}^{-1}[X(s)]$

(2) The solution for the output equations.
The output equation is given as $y(t) = Cx(t) + Df(t)$. Find $y(t)$.
Step 1: $F(s) = ?$
Step 2: $X(s) = \boldsymbol{\Phi}(s)x(0) + \boldsymbol{\Phi}(s)BF(s) = ?$
Step 3: $Y(s) = CX(s) + DF(s)$
Step 4: $y(t) = \mathcal{L}^{-1}[Y(s)]$

10.4.2 Solutions in the time domain

The matrix exponential e^{At} will be introduced as a new concept here. It can be defined by means of an infinite series as follows

$$e^{At} = I + At + \frac{A^2 t^2}{2} + \frac{A^3 t^3}{3!} + \cdots + \frac{A^n t^n}{n!} + \cdots = \sum_{k=1}^{\infty} \frac{A^k t^k}{k!} \qquad (10.4\text{-}13)$$

For example,

$$I = \begin{bmatrix} 1 & 0 \\ 0 & 1 \end{bmatrix} \quad \text{and} \quad A = \begin{bmatrix} 0 & 1 \\ 2 & 1 \end{bmatrix}, \quad \text{so} \quad At = \begin{bmatrix} 0 & 1 \\ 2 & 1 \end{bmatrix} t = \begin{bmatrix} 0 & t \\ 2t & t \end{bmatrix},$$

$$\frac{A^2 t^2}{2!} = \begin{bmatrix} 0 & 1 \\ 2 & 1 \end{bmatrix}\begin{bmatrix} 0 & 1 \\ 2 & 1 \end{bmatrix}\frac{t^2}{2} = \begin{bmatrix} 2 & 1 \\ 2 & 3 \end{bmatrix}\frac{t^2}{2} = \begin{bmatrix} t^2 & \frac{t^2}{2} \\ t^2 & \frac{3t^2}{2} \end{bmatrix}, \quad \text{etc.}$$

It can be proved that the infinite series of equation (10.4-13) are absolute and uniformly convergent for any time t. So, the series can be differentiated or integrated term by term. The differential form of equation (10.4-13) is of the form

$$\frac{d}{dt} e^{At} = A + A^2 t + \frac{A^3 t^2}{2!} + \cdots = A\left[I + At + \frac{A^2 t^2}{2!} + \frac{A^3 t^3}{3!} + \cdots\right] = A e^{At}.$$

Note: The expression can also be written as

$$\frac{d}{dt} e^{At} = A + A^2 t + \frac{A^3 t^2}{2!} + \cdots = \left[I + At + \frac{A^2 t^2}{2!} + \frac{A^3 t^3}{3!} + \cdots\right] A = e^{At} A.$$

That is,

$$\frac{d}{dt} e^{At} = A e^{At} = e^{At} A. \qquad (10.4\text{-}14)$$

In addition, from equation (10.4-13),

$$e^0 = I,$$

and then,

$$e^{-At} e^{At} = e^{At} e^{-At} = I. \qquad (10.4\text{-}15)$$

Next, we give the solution process for the state equation

$$\dot{x}(t) = Ax(t) + Bf(t). \qquad (10.4\text{-}16)$$

It can be proved that $\frac{d}{dt} UV = \frac{dU}{dt} V + U\frac{dV}{dt}$, where U and V are general variable matrixes without any physical meaning. Therefore,

$$\frac{d}{dt}\left[e^{-At} x(t)\right] = \left[\frac{d}{dt} e^{-At}\right] x(t) + e^{-At}\dot{x}(t) = -e^{-At} Ax(t) + e^{-At}\dot{x}(t). \qquad (10.4\text{-}17)$$

Premultiplying both sides of equation (10.4-16) by e^{-At}, we have

$$e^{-At}\dot{x}(t) = e^{-At}Ax(t) + e^{-At}Bf(t)$$

or

$$e^{-At}\dot{x}(t) - e^{-At}Ax(t) = e^{-At}Bf(t) . \qquad (10.4\text{-}18)$$

Comparing this with equation (10.4-17), the left side of equation (10.4-18) is just $\frac{d}{dt}[e^{-At}x(t)]$, and therefore,

$$\frac{d}{dt}[e^{-At}x(t)] = e^{-At}Bf(t) . \qquad (10.4\text{-}19)$$

Integrating both sides of equation (10.4-19) from 0 to t, we obtain

$$[e^{-At}x(t)]\Big|_0^t = \int_0^t e^{-A\tau}Bf(\tau)d\tau ,$$

$$e^{-At}x(t) = x(0) + \int_0^t e^{-A\tau}Bf(\tau)d\tau . \qquad (10.4\text{-}20)$$

After premultiplying both sides of equation (10.4-20) by e^{At} and using equation (10.4-15), the result is

$$x(t) = \underbrace{e^{At}x(0)}_{\text{zero input} \atop \text{component}} + \underbrace{\int_0^t e^{A(t-\tau)}Bf(\tau)d\tau}_{\text{zero state} \atop \text{component}} . \qquad (10.4\text{-}21)$$

This result conforms to the response decomposition introduced in previous chapters. Obviously, the second term in equation (10.4-21) is a convolution, so this equation can be also written as

$$x(t) = e^{At}x(0) + e^{At} * Bf(t) . \qquad (10.4\text{-}22)$$

Now we will give the solution methods for the output equation vector in the time domain. If the output equation vector is

$$y(t) = Cx(t) + Df(t) ,$$

substituting equation (10.4-22) into the above equation, we obtain

$$y(t) = C[e^{At}x(0) + e^{At} * Bf(t)] + Df(t) . \qquad (10.4\text{-}23)$$

The elements in B are all constants, so,

$$e^{At} * Bf(t) = e^{At}B * f(t) .$$

With this result, equation (10.4-23) can be changed into

$$y(t) = C[e^{At}x(0) + e^{At}B * f(t)] + Df(t) . \qquad (10.4\text{-}24)$$

As we know, a convolution of any signal and the impulse signal is still this signal, that is,

$$f(t) * \delta(t) = \delta(t) * f(t) = f(t) .$$

This result can be generalized into the matrix operation. Defining an $m \times m$-order diagonal matrix

$$\boldsymbol{\delta}(t) = \begin{bmatrix} \delta(t) & 0 & \cdots & 0 \\ 0 & \delta(t) & 0 & \vdots \\ \vdots & 0 & \ddots & 0 \\ 0 & \cdots & 0 & \delta(t) \end{bmatrix} , \tag{10.4-25}$$

we have

$$\boldsymbol{f}(t) * \boldsymbol{\delta}(t) = \boldsymbol{\delta}(t) * \boldsymbol{f}(t) = \boldsymbol{f}(t) .$$

Thus, equation (10.4-24) can be rewritten as

$$y(t) = C[e^{At}x(0) + e^{At}B * f(t)] + D\delta(t) * f(t)$$
$$= \underbrace{Ce^{At}x(0)}_{\text{zero input} \atop \text{component}} + \underbrace{[Ce^{At}B + D\delta(t)] * f(t)}_{\text{zero state} \atop \text{component}} . \tag{10.4-26}$$

If we set $\boldsymbol{\Phi}(t) = e^{At}$, equation (10.4-26) can be rewritten as

$$y(t) = \underbrace{C\boldsymbol{\Phi}(t)x(0)}_{\text{zero input} \atop \text{component}} + \underbrace{[C\boldsymbol{\Phi}(t)B + D\delta(t)] * f(t)}_{\text{zero state} \atop \text{component}} .$$

From the relationship between the zero-state response and the excitation introduced previously, we have

$$y_f(t) = \underbrace{[C\boldsymbol{\Phi}(t)B + D\delta(t)] * f(t)}_{\text{zero state} \atop \text{component}} = h(t) * f(t) , \tag{10.4-27}$$

where

$$h(t) = C\boldsymbol{\Phi}(t)B + D\delta(t) . \tag{10.4-28}$$

Thus,

$$y(t) = \underbrace{C\boldsymbol{\Phi}(t)x(0)}_{\text{zero input} \atop \text{component}} + \underbrace{h(t) * f(t)}_{\text{zero state} \atop \text{component}} . \tag{10.4-29}$$

The $\boldsymbol{h}(t)$ is called the impulse response matrix and is a $k \times m$-order matrix, where the element $h_{ij}(t)$ located in the ith row and the jth column represents the zero-state response $y_{fi}(t)$ when the input is $f_j(t) = \delta(t)$ but the other inputs are all zero.
    Note: The $\boldsymbol{h}(t)$ and $\boldsymbol{H}(s)$ can satisfy the Laplace transform relation

$$h(t) = \mathcal{L}^{-1}[H(s)] . \tag{10.4-30}$$

So far, steps of solving the state equations in the time domain are the following:

Step 1: Calculate $\boldsymbol{\Phi}(t) = e^{At}$.

Step 2: Find $\boldsymbol{x}(t) = e^{At}\boldsymbol{x}(0) + e^{At} * \boldsymbol{Bf}(t)$.

The steps of solving the output equations in the time domain are the following:

Step 1: Calculate $\boldsymbol{\Phi}(t) = e^{At}$.

Step 2: Find $\boldsymbol{h}(t) = \boldsymbol{C\Phi}(t)\boldsymbol{B} + \boldsymbol{D\delta}(t)$.

Step 3: Find $\boldsymbol{y}(t) = \underbrace{\boldsymbol{C\Phi}(t)\boldsymbol{x}(0)}_{\substack{\text{zero input} \\ \text{component}}} + \underbrace{\boldsymbol{h}(t) * \boldsymbol{f}(t)}_{\substack{\text{zero state} \\ \text{component}}}$.

10.4.3 Calculation of e^{At}

From the previous description, we know that e^{At} plays an important role in solving for the state model, but usually we cannot use equation (10.4-13) to calculate e^{At}, because the calculation is more complicated, and the result is an infinite series but not a closed expression. So, we need to find a good solution method for this problem.

Putting equations (10.4-5) and (10.4-21) together, we have

$$\boldsymbol{x}(t) = \underbrace{\mathcal{L}^{-1}[\boldsymbol{\Phi}(s)]\boldsymbol{x}(0)}_{\substack{\text{zero input} \\ \text{component}}} + \underbrace{\mathcal{L}^{-1}[\boldsymbol{\Phi}(s)\boldsymbol{BF}(s)]}_{\substack{\text{zero state} \\ \text{component}}},$$

$$\boldsymbol{x}(t) = \underbrace{e^{At}\boldsymbol{x}(0)}_{\substack{\text{zero input} \\ \text{component}}} + \underbrace{\int_0^t e^{A(t-\tau)}\boldsymbol{Bf}(\tau)\mathrm{d}\tau}_{\substack{\text{zero state} \\ \text{component}}}.$$

We find that

$$e^{At} = \mathcal{L}^{-1}[\boldsymbol{\Phi}(s)]. \qquad (10.4\text{-}31)$$

Namely, e^{At} and $\boldsymbol{\Phi}(s)$ are the Laplace transform pair, which yields

$$\boldsymbol{\Phi}(t) = e^{At} \overset{\mathcal{L}}{\longleftrightarrow} \boldsymbol{\Phi}(s). \qquad (10.4\text{-}32)$$

Equation (10.4-32) is just a shortcut to find e^{At} using the inverse Laplace transform.

The $\boldsymbol{\Phi}(t) = e^{At}$ is called the transition or transfer matrix and has the following features:

(1) $\boldsymbol{\Phi}(0) = \boldsymbol{I}$

(2) $\boldsymbol{\Phi}(t - t_0) = \boldsymbol{\Phi}(t - t_1)\boldsymbol{\Phi}(t_1 - t_0)$

(3) $\boldsymbol{\Phi}(t_1 + t_2) = \boldsymbol{\Phi}(t_1)\boldsymbol{\Phi}(t_2)$

(4) $[\boldsymbol{\Phi}(t)]^{-1} = \boldsymbol{\Phi}(-t)$

(5) $[\boldsymbol{\Phi}(t)]^n = \boldsymbol{\Phi}(nt)$

(6) $\frac{\mathrm{d}}{\mathrm{d}t}e^{At} = Ae^{At} = e^{At}A$

Example 10.4-4. For the system shown in Example 10.4-2, solve impulse response matrix of the system $\boldsymbol{h}(t)$ by the time domain method.

Solution. From Example 10.4-2, we have

$$\boldsymbol{\Phi}(s) = (s\boldsymbol{I} - \boldsymbol{A})^{-1} = \begin{bmatrix} \frac{s+3}{(s+1)(s+2)} & \frac{1}{(s+1)(s+2)} \\ \frac{-2}{(s+1)(s+2)} & \frac{s}{(s+1)(s+2)} \end{bmatrix}.$$

With the equation $\boldsymbol{\Phi}(t) = e^{At} = \mathcal{L}^{-1}[\boldsymbol{\Phi}(s)]$, we can obtain

$$\boldsymbol{\Phi}(t) = \mathcal{L}^{-1}\begin{bmatrix} \frac{s+3}{(s+1)(s+2)} & \frac{1}{(s+1)(s+2)} \\ \frac{-2}{(s+1)(s+2)} & \frac{s}{(s+1)(s+2)} \end{bmatrix} = \mathcal{L}^{-1}\begin{bmatrix} \frac{2}{s+1} - \frac{1}{s+2} & \frac{1}{s+1} - \frac{1}{s+2} \\ \frac{-2}{s+1} + \frac{2}{s+2} & \frac{-1}{s+1} - \frac{2}{s+2} \end{bmatrix}$$

$$= \begin{bmatrix} 2e^{-t} - e^{-2t} & e^{-t} - e^{-2t} \\ -2e^{-t} + 2e^{-2t} & -e^{-t} + 2e^{-2t} \end{bmatrix}.$$

With equation (10.4-25), we can obtain

$$\boldsymbol{\delta}(t) = \begin{bmatrix} \delta(t) & 0 \\ 0 & \delta(t) \end{bmatrix}.$$

With equation (10.4-28) we can obtain

$$\boldsymbol{h}(t) = \begin{bmatrix} 1 & 0 \\ 1 & 1 \\ 0 & 2 \end{bmatrix} \begin{bmatrix} 2e^{-t} - e^{-2t} & e^{-t} - e^{-2t} \\ -2e^{-t} + 2e^{-2t} & -e^{-t} + 2e^{-2t} \end{bmatrix} \begin{bmatrix} 1 & 0 \\ 1 & 1 \end{bmatrix} + \begin{bmatrix} 0 & 0 \\ 1 & 0 \\ 0 & 1 \end{bmatrix} \begin{bmatrix} \delta(t) & 0 \\ 0 & \delta(t) \end{bmatrix}$$

$$= \begin{bmatrix} 3e^{-t} - 2e^{-2t} & e^{-t} - e^{-2t} \\ \delta(t) + 2e^{-2t} & e^{-2t} \\ -6e^{-t} + 8e^{-2t} & \delta(t) - 2e^{-t} + 4e^{-2t} \end{bmatrix} \quad (t \geq 0).$$

Readers can verify that this result is just the inverse Laplace transform of $\boldsymbol{H}(s)$ in Example 10.4-2.

10.5 Judging stability

We know that the natural frequencies of the system are the roots of the characteristic equation $|s\boldsymbol{I} - \boldsymbol{A}| = 0$, namely the characteristic values. The related knowledge in Chapter 7 stated that when all the characteristic values of a system lie in the left half of the s plane, this system is stable. Therefore, once the roots of $|s\boldsymbol{I} - \boldsymbol{A}| = 0$ are found, the stability of the system can be judged by the Routh–Hurwitz Criterion.

Example 10.5-1. To make the system shown in ▸ Figure 10.10 stable, determine the range of gain k.

Solution. According to the system block diagram the state equations can be written as

$$\dot{x}_1 = x_2$$
$$\dot{x}_2 = -x_2 + k(-x_1 - x_3 + f_2)$$
$$\dot{x}_3 = -2x_3 + f_1 - (x_2 + x_3)$$

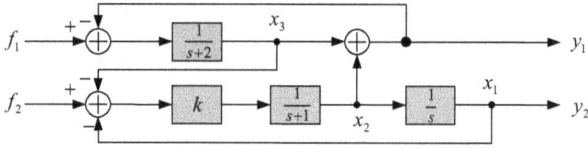

Fig. 10.10: E10.5-1.

In matrix from they can be written as

$$\begin{bmatrix} \dot{x}_1 \\ \dot{x}_2 \\ \dot{x}_3 \end{bmatrix} = \begin{bmatrix} 0 & 1 & 0 \\ -k & -1 & -k \\ 0 & -1 & -3 \end{bmatrix} \begin{bmatrix} x_1 \\ x_2 \\ x_3 \end{bmatrix} + \begin{bmatrix} 0 & 0 \\ 0 & k \\ 1 & 0 \end{bmatrix} \begin{bmatrix} f_1 \\ f_2 \end{bmatrix}$$

The characteristic polynomial is

$$|sI - A| = \begin{vmatrix} s & -1 & 0 \\ k & s+1 & k \\ 0 & 1 & s+3 \end{vmatrix} = s^3 + 4s^2 + 3s + 3k$$

The R-H array can be arranged as

$$\begin{array}{cc} 1 & 3 \\ 4 & 3k \\ \frac{12-3k}{4} & 0 \\ 3k & 0 \end{array}$$

Clearly, when $0 < k < 4$, the plus or minus signs of the elements in the first column are not changed; the system is stable.

10.6 Judging controllability and observability

From Section 7.3, we know that in a system described by the external method, because the output is both the observed and the controlled objective, and it directly relates to the input via the differential or the difference equation, there is no the controllability or observability of the system. However, in the state space description, the output is related to the input by the state variables, so the variation of the internal states affect the output; as a result, there will be two problems.

(1) Under the drive of the input, can a system change from the starting state to the required state in a limited time? This is the controllability problem.

(2) Can the starting state of a system be determined by observation of the output in a limited time interval? This is the observability problem.

The following conclusions can be deduced in this study. Suppose that M is a square matrix of nth-order (the number of the state variables is n). It can be defined as

$$M = \left(B \vdots AB \vdots A^2 B \vdots \cdots \vdots A^{n-2} B \vdots A^{n-1} B \right), \tag{10.6-1}$$

then M is nonsingular or full rank, which is the necessary and sufficient condition to test the controllability of a system. It is suitable for both discrete and continuous systems.

Suppose that N is a square matrix of nth-order (the number of the state variables is n). It can be defined as

$$N = \begin{bmatrix} C \\ \cdots \\ CA \\ \cdots \\ CA^2 \\ \cdots \\ \vdots \\ \cdots \\ CA^{n-1} \end{bmatrix}, \tag{10.6-2}$$

and then N is nonsingular or full rank, which is the necessary and sufficient condition to test the observability of a system. It is suitable for both discrete and continuous systems.

Although research on the controllability and observability of a system is an important application of the state space analysis method in the system analysis field, details will not be introduced in this book due to space limitations.

10.7 Solved questions

Question 10-1. The flow graph of a system is shown in ▸ Figure Q10-1. Write the state equations and output equations.

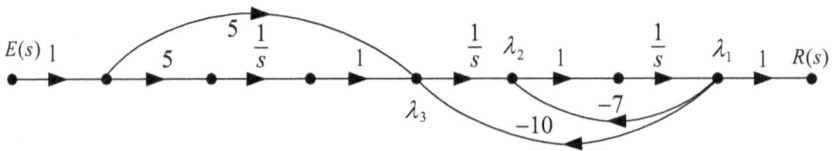

Fig. Q10-1

Solution. Based on the flow graph, the state equations can be obtained directly by

$$\begin{cases} \dot{\lambda}_1(t) = -7\lambda_1(t) + \lambda_2(t) \\ \dot{\lambda}_2(t) = -10\lambda_1(t) + \lambda_3(t) + 5e(t) \\ \dot{\lambda}_3(t) = 5e(t) \end{cases}.$$

So, the output equation of the system is

$$r(t) = \lambda_1(t).$$

Question 10-2. The state model of a system and the starting state vector are

$$\begin{bmatrix} \dot{x}_1 \\ \dot{x}_2 \end{bmatrix} = \begin{bmatrix} 1 & 0 \\ 1 & -3 \end{bmatrix}\begin{bmatrix} x_1 \\ x_2 \end{bmatrix} + \begin{bmatrix} 1 \\ 0 \end{bmatrix}x, \quad y = \begin{bmatrix} -\frac{1}{4} & 1 \end{bmatrix}\begin{bmatrix} x_1 \\ x_2 \end{bmatrix}, \quad x(0_-) = \begin{bmatrix} x_1(0_-) \\ x_1(0_-) \end{bmatrix} = \begin{bmatrix} 1 \\ 2 \end{bmatrix}.$$

If the input $x(t) = \varepsilon(t)$, find the system function matrix and the output $y(t)$.

Solution. Since

$$\Phi(s) = (sI - A)^{-1} = \begin{bmatrix} s-1 & 0 \\ -1 & s+3 \end{bmatrix}^{-1} = \frac{1}{(s-1)(s+3)}\begin{bmatrix} s+3 & 0 \\ 1 & s-1 \end{bmatrix}$$

$$= \begin{bmatrix} \frac{1}{s-1} & 0 \\ \frac{1}{(s-1)(s+3)} & \frac{1}{(s+3)} \end{bmatrix}$$

the system function matrix is

$$H(s) = C\Phi(s)B + D = \begin{bmatrix} -\frac{1}{4} & 1 \end{bmatrix}\begin{bmatrix} \frac{1}{s-1} & 0 \\ \frac{1}{(s-1)(s+3)} & \frac{1}{(s+3)} \end{bmatrix}\begin{bmatrix} 1 \\ 0 \end{bmatrix} = \frac{-1}{4(s+3)}.$$

The zero-state response is

$$y_f(t) = \mathcal{L}^{-1}[H(s)X(s)] = \mathcal{L}^{-1}\left[\frac{-1}{4(s+3)}\frac{1}{s}\right] = \mathcal{L}^{-1}\left[\frac{1}{12}\left(\frac{1}{s+3} - \frac{1}{s}\right)\right]$$

$$= \frac{1}{12}(e^{-3t} - 1)\varepsilon(t).$$

The zero-input response is

$$y_x(t) = \mathcal{L}^{-1}[C\Phi(s)x(0)] = \mathcal{L}^{-1}\begin{bmatrix} -\frac{1}{4} & 1 \end{bmatrix}\begin{bmatrix} \frac{1}{s-1} & 0 \\ \frac{1}{(s-1)(s+3)} & \frac{1}{(s+3)} \end{bmatrix}\begin{bmatrix} 1 \\ 2 \end{bmatrix}$$

$$= \mathcal{L}^{-1}\left[\frac{7}{4}\frac{1}{s+3}\right] = \frac{7}{4}e^{-3t}\varepsilon(t)$$

So, the complete response is

$$y(t) = y_f(t) + y_x(t) = \frac{1}{12}\left(e^{-3t} - 1\right)\varepsilon(t) + \frac{7}{4}e^{-3t}\varepsilon(t) = \left(\frac{11}{6}e^{-3t} - \frac{1}{12}\right)\varepsilon(t)$$

Question 10-3. For the system is shown in ▶ Figure Q10-3 (1), $R = 1\,\Omega$, $C = 0.5\,\text{F}$, $u_C(0_-) = 1\,\text{V}$, $u_L(0_-) = 1\,\text{A}$, $u_S(t) = \varepsilon(t)\,\text{V}$, $i_S(t) = \varepsilon(t)\,\text{A}$, and $\varepsilon(t)$ is unit step function.
(1) Work out the equivalent circuit of this system in the s domain.
(2) Find the total response of $i_R(t)$ through resistor R.
(3) If $u_C(t) = x_1$, $i_L(t) = x_2$, try to build the state equations of this circuit.

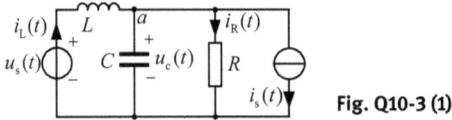

Fig. Q10-3 (1)

Solution. (1) The equivalent circuit in the s domain can be drawn as in ▶ Figure Q10-3 (2).

(2) Letting the voltage on the resistor R be $U_R(s)$, the node voltage equation is

$$\left(\frac{1}{sL} + sC + \frac{1}{R}\right) U_R(s) = \frac{1}{sL}\left[Li_L(0_-) + U_S(s)\right] + Cu_C(0_-) - I_S(s).$$

Applying the Laplace transform to $u_S(t) = \varepsilon(t)$ V and $i_S(t) = \varepsilon(t)$ A, respectively, we have $U_S(s) = \frac{1}{s}$ and $I_S(s) = \frac{1}{s}$.
Substituting the known conditions into the equation yields

$$\left(\frac{1}{s} + \frac{s}{2} + 1\right) U_R(s) = \frac{1}{s}\left(1 + \frac{1}{s}\right) + \frac{1}{2} - \frac{1}{s} = \frac{1}{s^2} + \frac{1}{2}.$$

So,

$$U_R(s) = \frac{s^2 + 2}{s(s^2 + 2s + 2)} = \frac{s^2 + 2 + 2s - 2s}{s(s^2 + 2s + 2)}$$

$$= \frac{1}{s} - \frac{2}{s^2 + 2s + 2} = \frac{1}{s} - \frac{2}{(s+1)^2 + 1}.$$

Applying the inverse Laplace transform to the equation yields

$$u_R(t) = \left(1 - 2e^{-t}\sin t\right)\varepsilon(t).$$

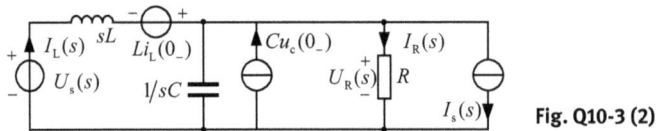

Fig. Q10-3 (2)

So, the total current response through the resistor R is

$$i_R(t) = u_R(t)/R = \left(1 - 2e^{-t}\sin t\right)\varepsilon(t) .$$

(3) The KCL equation on node a in ▶ Figure Q10-3 (1) is

$$C\dot{x}_1 + \frac{x_1}{R} + i_S - x_2 = 0 .$$

Substituting the known conditions into the equation, we have

$$\dot{x}_1 = -2x_1 + 2x_2 - 2i_S .$$

The KVL equation of the left loop can be written as

$$L\dot{x}_2 + x_1 - u_S = 0 .$$

Substituting the known condition into the equation, we have

$$\dot{x}_2 = -x_1 + u_S .$$

So, the state equations can be written in matrix form as

$$\begin{bmatrix} \dot{x}_1 \\ \dot{x}_2 \end{bmatrix} = \begin{bmatrix} -2 & 2 \\ -1 & 0 \end{bmatrix} \begin{bmatrix} x_1 \\ x_2 \end{bmatrix} + \begin{bmatrix} 0 & -2 \\ 1 & 0 \end{bmatrix} \begin{bmatrix} u_S \\ i_S \end{bmatrix}$$

10.8 Learning tips

The state model is another type of mathematical description of a system after the equation model, the impulse response and the system function. The state space analysis is a method based on the matrix operation. Readers should note the following points.

(1) The state space method is a way to analyze a system via seeking intermediate variables that can relate to both the excitations and the responses in the system. This view of the analysis problem is very valuable.

(2) The state variables not only are a bridge between the excitations and responses but can also reflect historical behaviors of a system and can determine the current and future values of the system with the current excitations.

(3) The knowledge from the circuit analysis and linear algebra courses is the basis of the state space method.

10.9 Problems

Problem 10-1. Write the state equations of the circuits shown in ▶ Figure P10-1.

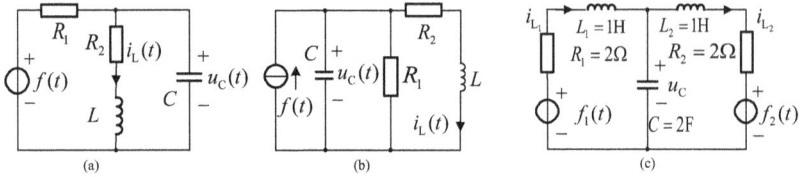

Fig. P10-1

Problem 10-2. For the complex system shown in ▶ Figure P10-2, write the state equations and the output equations.

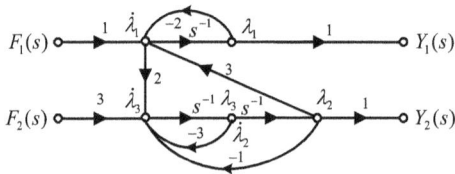

Fig. P10-2

Problem 10-3. Write the state models of the circuits which are shown in ▶ Figure P10-3.

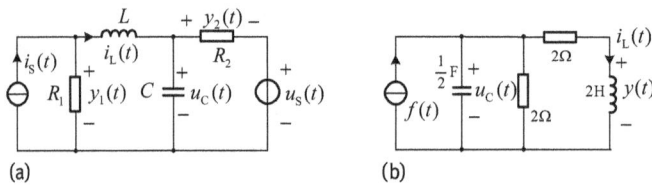

Fig. P10-3

Problem 10-4. Write the state models of the systems represented by the block diagrams.

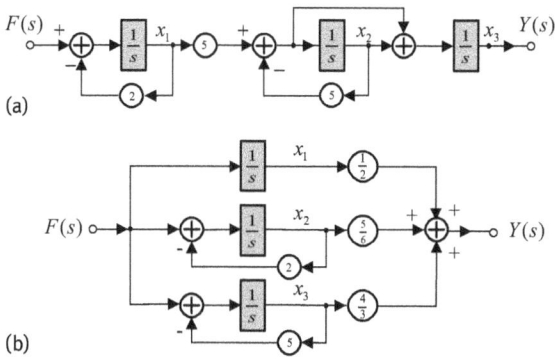

(a)

(b)   Fig. P10-4

Problem 10-5. A system is shown in ▶ Figure P10-5, $x_1(t)$, $x_2(t)$, $x_3(t)$ are the state variables. Write the state model of the system.

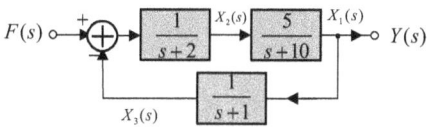

Fig. P10-5

Problem 10-6. Flow graphs of systems are shown in ▶ Figure P10-6. Write the state model of each system.

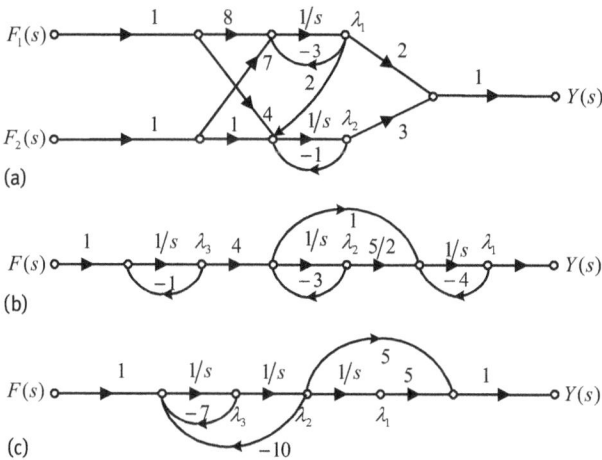

(a)

(b)

(c)   Fig. P10-6

Problem 10-7. Differential equations of systems are as follows. Write the state model of each system.

(1) $y''(t) + 4y'(t) + 3y(t) = f'(t) + f(t)$

(2) $y'''(t) + 5y''(t) + y'(t) + 2y(t) = f'(t) + 2f(t)$

Problem 10-8. $H(s) = \frac{3s+10}{s^2+7s+12}$ is known. Draw three forms of signal flow graph and write the corresponding state models.

Problem 10-9. Some A matrixes are as follows:

$$(1)\ A = \begin{bmatrix} 0 & 2 \\ -1 & -2 \end{bmatrix} \qquad (2)\ A = \begin{bmatrix} 0 & 1 & 0 \\ 0 & 0 & 1 \\ 0 & 1 & 0 \end{bmatrix} \qquad (3)\ A = \begin{bmatrix} -1 & 0 & 0 \\ 0 & 2 & 0 \\ 0 & 0 & -3 \end{bmatrix}$$

Find each system's state transfer matrix $\Phi(t)$ and natural frequencies and judge the stability of each system.

Problem 10-10. The state model of a system is

$$\begin{bmatrix} \dot{x}_1 \\ \dot{x}_2 \end{bmatrix} = \begin{bmatrix} -2 & 1 \\ 0 & -1 \end{bmatrix} \begin{bmatrix} x_1 \\ x_2 \end{bmatrix} + \begin{bmatrix} 1 \\ 0 \end{bmatrix} f(t), \quad [y(t)] = \begin{bmatrix} 1 & 0 \end{bmatrix} \begin{bmatrix} x_1 \\ x_2 \end{bmatrix},$$

the starting states are $x_1(0_-) = x_2(0_-) = 1$, and the input is $f(t) = \varepsilon(t)$. Find $y(t)$, $H(s)$ and $h(t)$.

Problem 10-11. The state transfer matrix of the system is $\Phi(t)$. Find the corresponding matrix A.

$$(1)\ \Phi(t) = \begin{bmatrix} e^{-t} & 0 & 0 \\ 0 & (1-2t)e^{-2t} & 4te^{-2t} \\ 0 & -te^{-2t} & (1+2t)e^{-2t} \end{bmatrix};$$

$$(2)\ \Phi(t) = \begin{bmatrix} (1+t)e^{-t} & te^{-t} \\ -te^{-t} & (1-t)e^{-t} \end{bmatrix}$$

Problem 10-12. The state and the output equations of an LTI system are

$$\dot{x}(t) = \begin{bmatrix} -2 & 0 & 0 \\ 5 & -5 & 0 \\ 5 & -4 & 0 \end{bmatrix} x(t) + \begin{bmatrix} 1 \\ 0 \\ 0 \end{bmatrix} f(t), \quad y(t) = \begin{bmatrix} 0 & 0 & 1 \end{bmatrix} x(t)$$

Find:

(1) the system function $H(s)$;   (2) the impulse response.

Problem 10-13. The system matrix parameters and the input and starting states are as follows. Find the system response.

$$A = \begin{bmatrix} 0 & 1 \\ -3 & -4 \end{bmatrix}, \quad B = \begin{bmatrix} 0 \\ 2 \end{bmatrix}, \quad C = \begin{bmatrix} -1 & -2 \end{bmatrix}, \quad D = 1, \quad f(t) = \varepsilon(t), \quad x(0_-) = \begin{bmatrix} 1 \\ 2 \end{bmatrix}$$

Problem 10-14. A continuous system is shown in ▶ Figure P10-14.
(1) Find the state equations of the system.
(2) According to the state equations, write the differential equation of the system.
(3) When the excitation $f(t) = \varepsilon(t)$, the response is $y(t) = \left(\frac{1}{3} + \frac{1}{2}e^{-t} - \frac{5}{6}e^{-3t}\right)\varepsilon(t)$. Find the starting state $x(0_-)$ of the system.

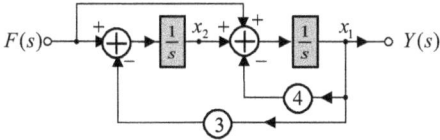

Fig. P10-14

Problem 10-15. There is an LTI system with zero input, $\dot{x}(t) = Ax(t)$. When

$$x(0_-) = \begin{bmatrix} 2 \\ 1 \end{bmatrix}, \quad x(t) = \begin{bmatrix} 6e^{-t} - 4e^{-2t} \\ -3e^{-t} + 4e^{-2t} \end{bmatrix};$$

when

$$x(0_-) = \begin{bmatrix} 0 \\ 1 \end{bmatrix}, \quad x(t) = \begin{bmatrix} 2e^{-t} - 2e^{-2t} \\ -e^{-t} + 2e^{-2t} \end{bmatrix}.$$

Find:
(1) the state transfer matrix $\boldsymbol{\Phi}(t)$;   (2) the corresponding matrix A.

Problem 10-16. Two simple systems are s_1 and s_2, as shown in ▶ Figure P10-16, and the state equations and the output equations are

system s_1: $\begin{cases} \dot{x}_1(t) = A_1x_1(t) + B_1f_1(t) \\ y_1(t) = C_1x_1(t) \end{cases}$, $A_1 = \begin{bmatrix} 0 & 1 \\ -2 & -3 \end{bmatrix}$, $B_1 = \begin{bmatrix} 0 \\ 1 \end{bmatrix}$, $C_1 = \begin{bmatrix} 3 & 1 \end{bmatrix}$

system s_2: $\begin{cases} \dot{x}_2(t) = A_2x_2(t) + B_2f_2(t) \\ y_2(t) = C_2x_2(t) \end{cases}$, $A_2 = -3$, $B_2 = 1$, $C_2 = 1$

Now consider two series systems, such as graphs (a) and (b), called the system.s_{12} and the system s_{21}. Find:
(1) the state models of system s_{12} and system s_{21};
(2) each transfer function of system s_1, s_2, s_{12} and s_{21}.

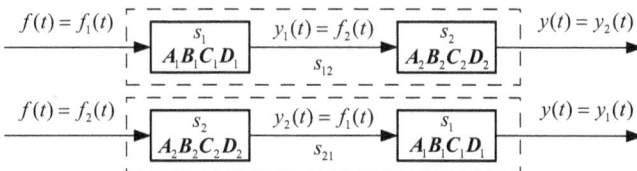

Fig. P10-16

Problem 10-17. The matrix parameters of a system are as follows. Find the system function $H(s)$.

$$A = \begin{bmatrix} 0 & 1 \\ -a_0 & -a_1 \end{bmatrix}, \quad B = \begin{bmatrix} 0 \\ 1 \end{bmatrix}, \quad C = \begin{bmatrix} b_0 - b_2 a_0 & b_1 - b_2 a_1 \end{bmatrix}, \quad D = b_2$$

Problem 10-18. The A matrix of a system is $A = \begin{bmatrix} 3 & -2 \\ 1 & 0 \end{bmatrix}$. Judge the system stability.

11 Applications of signals and systems analysis theories and methods

Questions: We have learned the basic theories and methods for signals and systems analysis from previous chapters. Then what are the practical applications of this knowledge and these skills?

Solutions: Take communication systems as examples and study the performance of several typical subsystems in them.

Results: A nondistortion transmission, equalizing, filtering, modulation and demodulation systems.

The preceding ten chapters have shown us some theories and solution methods for signals and systems analysis. In this chapter, we will use these theories and methods to analyze several typical subsystems in real communication systems. This will link theory with practice and lead to a satisfactory conclusion to the analysis course on signals and systems. At the same time, this chapter can be considered as a bridge between the "signals and systems" and "communication principles".

11.1 Nondistortion transmission systems

There are two possible results after an excitation signal is transmitted through a system.

(1) The waveform of the response is different from that of the excitation, that is, the distortions are produced by the system.

(2) The waveform of the response is similar to that of the excitation but is different with respect to the amplitude with a constant or different in time with time shifting or both. These cases are called nondistortion transmission.

For some signal processing systems, such as filters, modulators, frequency multipliers and so on, the existence of distortion is the requirement of the system, or it is necessary. However, the undistorted transmission of signals is the supreme goal pursued for a common amplifier. Therefore, a brief discussion will be given in this section about the condition of nondistortion transmission for a communication system.

According to the concept of nondistortion and the characteristics of an LTI system, the nondistortion response $y(t)$ generated by a continuous system to the excitation $f(t)$ should hold,

$$y(t) = Kf(t - t_d) . \tag{11.1-1}$$

In this expression, both the gain K and the time delay t_d are constant. This is considered as the nondistortion transmission condition of a system in the time domain.

https://doi.org/10.1515/9783110541205-004

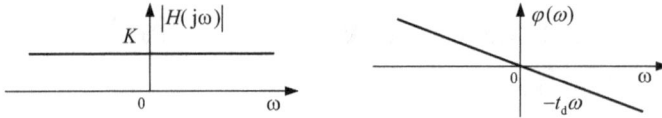

Fig. 11.1: The ideal nondistortion transmission system frequency characteristics.

Applying the Fourier transform to both sides of equation (11.1-1), we have

$$Y(j\omega) = KF(j\omega)e^{-j\omega t_\mathrm{d}} .\tag{11.1-2}$$

We know that the system function, the Fourier transforms of response and excitation, satisfy

$$Y(j\omega) = F(j\omega)H(j\omega) .\tag{11.1-3}$$

Comparing equations (11.1-2) and (11.1-3), we obtain

$$H(j\omega) = Ke^{-j\omega t_\mathrm{d}} .\tag{11.1-4}$$

Equation (11.1-4) is called the nondistortion transmission condition of a system in the frequency domain. This means that if a system is going to transmit a signal with nondistortion, the amplitude spectrum of the system function $|H(j\omega)|$ should be a constant K and the phase spectrum $\varphi(\omega)$ should be a line $-t_\mathrm{d}\omega$ passing through the origin; these are plotted in ▶ Figure 11.1.

Although it is impossible to achieve the ideal nondistortion transmission system in reality, giving the nondistortion transmission conditions still has theoretical significance.

Physical signals in the actual communication process all are limited in bandwidth, so they can be called band limited signals. For example, the band of the human voice signal on the phone ranges from 300 Hz to 3400 Hz, while that of the analog video signal ranges from 0 MHz to 6 MHz. Simply speaking, a signal with the up limited frequency ω_H is a band limited signal. Obviously, in order to transfer it, the frequency characteristics of a system should only be assured to satisfy the nondistortion condition in a band that is no less than the frequency spectrum of the signal, namely,

$$H(j\omega) = Ke^{-j\omega t_\mathrm{d}} \qquad |\omega| < \omega_\mathrm{H} .\tag{11.1-5}$$

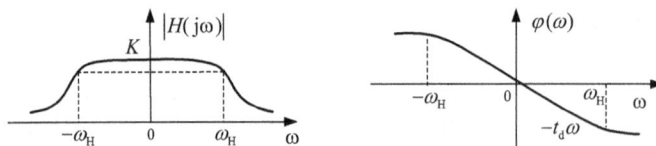

Fig. 11.2: Band limited signal nondistortion transmission system frequency characteristics.

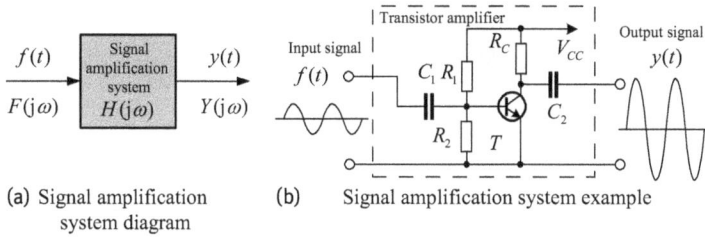

(a) Signal amplification   (b)   Signal amplification system example
system diagram

Fig. 11.3: Signal amplification system schematic diagram and example.

Equation (11.1-5) is also known as the band limited nondistortion transmission condition; $|H(j\omega)|$ and $\varphi(\omega)$ are sketched in ► Figure 11.2. In general, the frequency range $|\omega| < \omega_H$ in the frequency characteristics of a system is called the passband, which means the frequency range for nondistortion transmission or, simply, B. The ω_H is usually the corresponding to frequency value, which is a value of the amplitude spectrum going down to 70% of its maximum value, namely, $|H(j\omega_H)| = 70K\%$.

In the passband B, if the amplitude spectrum $|H(j\omega)|$ is not a constant, the transmitted signal will suffer from the amplitude frequency distortions. However, if the waveform of the phase spectrum $\varphi(\omega)$ is not a straight line, the transmitted signal will be subjected to phase frequency distortions. No matter which circumstance appears or if both occur at the same time, the output waveform will be distorted, which indicates that the system is not yet a nondistortion transmission system.

If $K > 1$ in equation (11.1-5), the system with this frequency characteristic should be a signal amplifier. A typical amplifier block and its schematic diagrams are shown in ► Figure 11.3.

If $0 < K < 1$ in equation (11.1-5), the system with this frequency characteristic is a signal attenuator.

11.2 Equilibrium systems

For a practical communication system, there is inevitably a certain degree of distortions to a signal which is transmitted by it. Therefore, in practice, the vital factor is how much the tolerance humans have for distortions. For example, because the human ear is sensitive to amplitude frequency distortions but is not sensitive to phase frequency distortions, for audio signal transmission or processing systems, people will have higher requirements for the flatness of the amplitude frequency characteristic $|H(j\omega)|$ and relatively lower requirements for the linearity of the phase frequency characteristic $\varphi(\omega)$. Again, for example, because the human visual system is not sensitive to amplitude frequency distortions of an image signal and finds it is difficult to accept its phase distortions, it is necessary to strictly control the nonlinear distortions of the phase characteristic in a video signal transmission system. For data signals, both am-

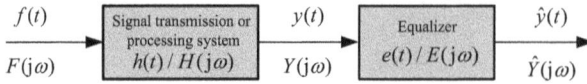

Fig. 11.4: Signal equilibrium system model diagram.

plitude frequency and phase frequency distortions of a system will have a great impact on the bit error rate of the signal transmission, so the requirements are all relatively higher.

If the distortions of a communication system are too large and go beyond the tolerance scope of people, then this problem can be solved by the equilibrium technique. In addition, if the frequency characteristics of a communication system do not meet people's requirements, for example, a music lover feels that the bass is from his audio amplifying system is not strong enough, that is, the low frequency segment of the amplitude frequency characteristic curve is not high enough, then the low frequency section's curve must be improved by a balancing technique to enhance the output power of the bass.

Equilibrium refers to a method that can compensate or correct the frequency characteristics for the whole system by controlling or changing the relative size of amplitudes or phases of signals with different frequencies or different frequency components in one signal.

Circuits, devices and systems that can realize this function are called equalizers, which are often placed behind a signal transmission or processing system to compensate or correct the system frequency characteristics and to meet the people's requirements for signal transmission.

If the impulse response of a signal transmission system is $h(t)$, the impulse response of an equalizer is $e(t)$. According to the nondistortion requirement, the time domain and frequency domain conditions of an equalization system (see ▶ Figure 11.4) must separately be

$$h(t) * e(t) = K\delta(t - t_\mathrm{d}) , \tag{11.2-1}$$

$$H(j\omega)E(j\omega) = Ke^{-j\omega t_\mathrm{d}} . \tag{11.2-2}$$

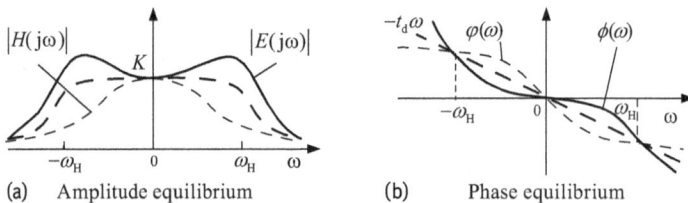

Fig. 11.5: Equilibrium system frequency characteristic.

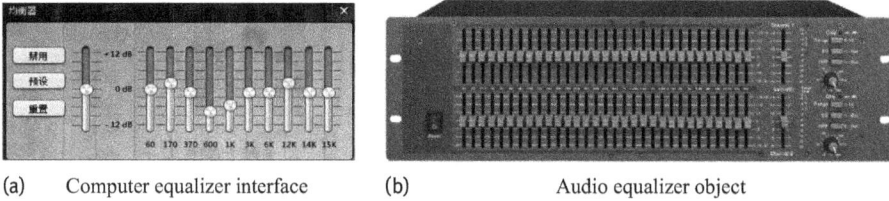

(a)   Computer equalizer interface   (b)   Audio equalizer object

Fig. 11.6: Equalizer instance diagram.

From equation (11.2-2), the frequency characteristics of the equalizer are

$$\begin{cases} |E(j\omega)| = \frac{K}{|H(j\omega)|} \\ \phi(\omega) = -\varphi(\omega) - \omega t_{\mathrm{d}} \end{cases} \quad , \quad |\omega| < \omega_{\mathrm{H}} \tag{11.2-3}$$

Obviously, based on the different circumstances to which it is applied, the equalizer can only compensate for amplitude distortions, which is called amplitude equilibrium, and the corresponding system is the amplitude equalizer. It can also only compensate for phase distortions, which is called the phase equilibrium and the corresponding system is the phase equalizer. ▸ Figure 11.5 shows the magnitude and phase equilibrium schematic diagrams; thick broken lines in the figure show the frequency characteristics that have been compensated by the equalizer.

An application example of amplitude equalization is the audio equalizer. ▸ Figure 11.6 shows two examples of an audio equalizer, each push rod controls the amplitude frequency characteristic of a narrow bandpass filter; if all pushrods are set at the same high level, the total amplitude frequency curve should be a flat topped curve of a broadband bandpass filter. If you want to upgrade (attenuation) a frequency component in a signal, then push up (down) the corresponding push rod.

The above equilibrium method is a kind of frequency domain equalization based on equation (11.2-2). In fact, from equation (11.2-1) there is also a time domain equalization method. Due to space limitations, however, this method is not introduced in this book.

11.3 Filtering systems

The most important and wide application of an LTI system is to filter signals.

Filtering is a signal processing method to select or restrain some alternating signals with different frequencies from different demands of people. The circuits, devices or systems that can achieve this function are called filters.

There are some extensions and methods of the filtering concept; some examples are given below.

(1) Selective filtering. A frequency selective filter is usually used to extract useful signals from interference signals. Its principle is to allow useful signals with a cer-

tain frequency or frequency band to pass through the system, while inhibiting or filtering useless signals or interference signals in other frequencies or frequency bands. A well known example is the tuner in radios or TV sets.

(2) Edge enhancement and contour extraction. For example, in image processing, in order to highlight the edge or contour of the object, a filter can be used to enhance the change rate of object edges and to facilitate object recognition or to extract contours from the image.

(3) Noise reduction. In all signal transmission and processing systems, the influence of noise and interference will appear throughout the whole process of signal acquisition, transmission and processing, such as snowflakes on the TV screen, the current sounds and background noises from a radio loudspeaker, and so on. Therefore, for a practical communication system, an important indicator in the design and evaluation is anti-noise performance. The selective filter is usually used to restrain noise or disturbing signals outside of the wanted signal frequency band.

(4) Equilibrium. Equilibrium is also considered as a type of filtering technology. The purpose of making up non-ideal frequency characteristics of a transmission system can be achieved by changing the relative sizes of magnitude and phase of each frequency component in a signal. Also, the equalizer is sometimes called the frequency shaping filter, which can change the spectrum shape.

Filtering via an LTI filter is called linear filtering. Of course, there is also the concept of nonlinear filtering, which is not discussed in this book.

The filter is usually divided into two kinds, i.e. ideal filters and actual filters, which are discussed herein. The ideal filter is one which the signals in its passband can pass through without any distortion, whereas signals out of the passband (in the stopband) are completely inhibited or eliminated, and there is no transitional zone between the passband and the stopband. In contrast, the actual filter is such a filter of which the signals in its passband can pass through it with a certain degree of distortion, and the signal outside the passband cannot be completely inhibited or filtered, and has a clear transitional zone between the passband and the stopband.

Several commonly used filters such as lowpass, highpass, bandpass and band stop filters are introduced in the following.

11.3.1 Ideal filters

(1) The amplitude frequency characteristic of an ideal lowpass filter is of the form

$$|H_{LP}(j\omega)| = \begin{cases} 1, & |\omega| < \omega_C \\ 0, & |\omega| > \omega_C \end{cases}. \tag{11.3-1}$$

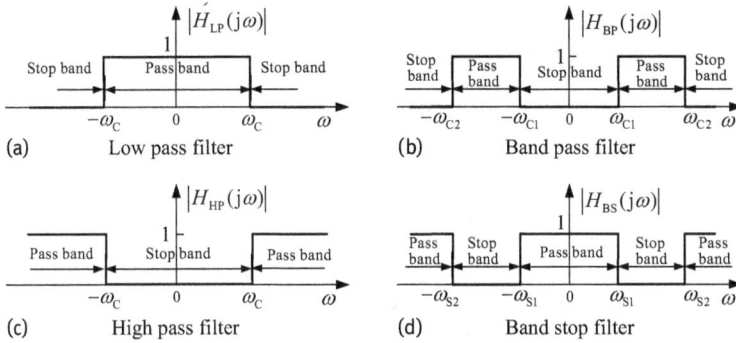

Fig. 11.7: The amplitude frequency characteristics of the ideal filters.

(2) The amplitude frequency characteristic of an ideal highpass filter is of the form

$$|H_{HP}(j\omega)| = \begin{cases} 1, & |\omega| > \omega_C \\ 0, & |\omega| < \omega_C \end{cases}.$$ (11.3-2)

(3) The amplitude frequency characteristic of an ideal bandpass filter is of the form

$$|H_{BP}(j\omega)| = \begin{cases} 1, & \omega_{C1} < |\omega| < \omega_{C2} \\ 0, & \text{other} \end{cases}.$$ (11.3-3)

(4) The amplitude frequency characteristic of an ideal bandstop filter is of the form

$$|H_{BS}(j\omega)| = \begin{cases} 0, & \omega_{S1} < |\omega| < \omega_{S2} \\ 1, & \text{other} \end{cases}.$$ (11.3-4)

In these expressions, ω_C is called the passband cutoff frequency, ω_{C1} is called the lower cutoff frequency, ω_{C2} is called the upper cutoff frequency, $\omega_{C1} < \omega_{C2}$, ω_{S1} is called the lower stopband frequency, ω_{S2} is called the upper stopband frequency and $\omega_{S1} < \omega_{S2}$.

The amplitude frequency characteristics of the four filters are shown in ▶ Figure 11.7.

Example 11.3-1. The frequency characteristic of an ideal highpass filter is

$$H_{HP}(j\omega) = \begin{cases} 1e^{-j\omega t_0}, & |\omega| > \omega_C \\ 0, & |\omega| < \omega_C \end{cases}$$

Solve the impulse response $h(t)$.

Solution. A gate signal $G_{2\omega_C}(\omega)$ can be introduced in the frequency domain, so the highpass filter's amplitude frequency characteristic can be written as

$$|H_{HP}(j\omega)| = 1 - G_{2\omega_C}(\omega).$$

Then, because

$$\delta(t) - \frac{\omega_C}{\pi} Sa\,(\omega_C t) \overset{\mathcal{F}}{\longleftrightarrow} 1 - G_{2\omega_C}(\omega),$$

we have

$$h(t) = \mathcal{F}^{-1}[H_{HP}(j\omega)] = \mathcal{F}^{-1}\Big[|H_{HP}(j\omega)|\,e^{-j\omega t_0}\Big] = \mathcal{F}^{-1}\Big[(1 - G_{2\omega_C}(\omega))\,e^{-j\omega t_0}\Big]$$

$$= \delta\,(t - t_0) - \frac{\omega_C}{\pi} Sa\,[\omega_C(t - t_0)]\,.$$

Example 11.3-2. A right sided exponential signal $f(t) = 2e^{-0.2t}\varepsilon(t)$ passes through an ideal lowpass filter. If the energy of the output signal is at least equal to 50% of the input signal energy, solve the cutoff frequency ω_C of the filter.

Solution. The spectrum of $f(t)$ can be obtained by equation (5.2-4)

$$F(j\omega) = \frac{2}{0.2 + j\omega}\,. \tag{11.3-5}$$

The energy of $f(t)$ is obtained by equation (5.3-19)

$$E_f = \int_{-\infty}^{+\infty} |f(t)|^2\,dt = \frac{1}{2\pi} \int_{-\infty}^{+\infty} |F(j\omega)|^2\,d\omega = \frac{1}{\pi} \int_{0}^{+\infty} \left(\frac{2}{0.2 + j\omega}\right)^2 d\omega \tag{11.3-6}$$

$$= \frac{1}{\pi} \int_{0}^{+\infty} \frac{4}{0.2^2 + \omega^2}\,d\omega = \frac{4}{\pi}[5\arctan(5\omega)]_0^{\infty} = 10\,J\,.$$

The spectrum of the output signal is obtained by equation (5.6-5)

$$Y(j\omega) = F(j\omega)H(j\omega) = \frac{2}{0.2 + j\omega} \qquad |\omega| \le \omega_C\,. \tag{11.3-7}$$

Then, the energy of the output signal is

$$E_y = \frac{1}{2\pi} \int_{-\infty}^{+\infty} |Y(j\omega)|^2\,d\omega = \frac{1}{\pi} \int_{0}^{\omega_C} \left(\frac{2}{0.2 + j\omega}\right)^2 d\omega = \frac{4}{\pi} \cdot 5\arctan(5\omega_C)\,. \tag{11.3-8}$$

Because the problem is

$$E_y = \frac{1}{2}E_f\,,$$

we have

$$\frac{4}{\pi} \cdot 5\arctan(5\omega_C) = 5\,. \tag{11.3-9}$$

From the relation above, the answer is

$$\omega_C = 0.2\,\mathrm{rad/s}\,.$$

Because the ideal filter has the same steep amplitude frequency characteristic waveform as a rectangle, which also is called the rectangle characteristic (it cannot be realized in practice), the study of the ideal filter is only of theoretical significance. Therefore, we also need to understand the knowledge about actual filter that can be realized in practice.

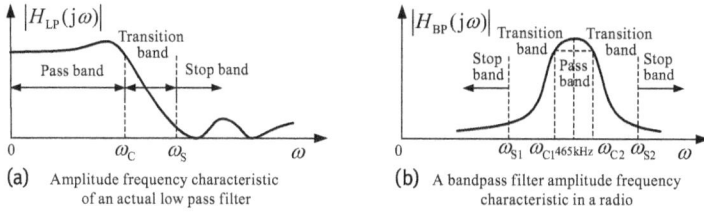

(a) Amplitude frequency characteristic
of an actual low pass filter

(b) A bandpass filter amplitude frequency
characteristic in a radio

Fig. 11.8: Amplitude frequency characteristics of actual filters.

11.3.2 Real filters

A realizable filter in practice cannot have the rectangle feature and its passband is generally not a flat top but a slight fluctuation curve, and its stopband is not always zero but may be a vibration attenuation curve. In addition, the spectrum of an actual filter is not double sided but only right sided.

The amplitude frequency characteristics of real lowpass and bandpass filters are shown in ▸ Figure 11.8, where ω_S is the stopband boundary frequency. It can be seen that the top of a real filter is not flat and the edge is not steep. So, whether the performance of a filter is good or bad, to a large extent depends on the steep degree of the frequency characteristic edge. The steeper the edge, the narrower the transition zone, the better the filtering performance, but the more complex the filter, the higher the cost. The steep degree of two edges of the characteristic curve of a filter relates to the order of the filter (system). Namely, the higher the order of a filter, the closer the characteristic curve is to the rectangle.

Systems such as the nondistortion transmission system, the equalization system and the filtering system all take the continuous system as the example, but the concepts and methods are also suitable for the study of discrete systems. This means that there are also concepts and systems called digital nondistortion, digital equalization, digital filtering and so on. For more details, the interested reader is referred to other books.

11.4 Modulation/demodulation systems

The modulation property of the Fourier transform is very useful in communication systems, which is the basic theory of wireless transmission and frequency division multiplexing. *Modulation refers to a process or a method of which a signal controls a certain parameter of another signal.*

In modulation techniques, the controlling signal is usually called the *modulating signal* or the original signal, which is a lower frequency signal. The controlled signal is called the *carrier*, which is a higher frequency periodic signal and does not contain useful information. In analog and digital modulation systems, the common carrier is a sinusoidal signal. After it is modulated by the modulating signal the carrier signal changes to a higher frequency signal carrying useful information and is called the *modulated signal.*

The signal processing procedure or method by which the modulating signal is taken from the modulated signal is called demodulation.

A real life example can help us to understand the concept of modulation/demodulation. If we want to carry goods to a place thousands of kilometers away, a truck, a train, a plane or a ship must be employed. Here, the goods is equivalent to the modulating signal, the vehicle is equivalent to the carrier. The procedure by which the goods are loaded onto a vehicle is equivalent to the modulation. The vehicle carrying the goods is the modulated signal. The unloading cargo process from the vehicle is the demodulation. This example may not be very appropriate, but it is substantially similar to the modulation and demodulation principles.

The main purposes of the modulation to a signal are to

(1) change a low frequency signal into a high frequency one to realize radio communication;

(2) improve the transmission quality of a signal;

(3) improve the communication efficiency and capacity of a system by using frequency division multiplexing technology.

Based on different controlled parameters in a carrier, three kinds of modulation systems are usually used in analog communication technology. These are called, respectively, amplitude modulation, frequency modulation and phase modulation.

Next, we will give a brief introduction to two kinds of common amplitude modulation techniques (systems), which are the suppressed carrier double sideband amplitude modulation (DSB) and the conventional double sideband amplitude modulation (AM).

11.4.1 DSB modulation system

If a useful signal for the transmission is $f(t)$, the carrier is a sinusoidal signal $\cos \omega_C t$, and $s_{DSB}(t)$ is a modulated signal. Now, the signal and the carrier are multiplied (the process is plotted in ▶ Figure 11.9), and we have

$$s_{DSB}(t) = f(t) \cos \omega_C t .\qquad(11.4\text{-}1)$$

According to the modulation property of the Fourier transform,

$$f(t) \overset{\mathcal{F}}{\longleftrightarrow} F(\omega)$$

$$\cos \omega_C t \overset{\mathcal{F}}{\longleftrightarrow} \pi [\delta (\omega + \omega_C) + \delta (\omega - \omega_C)]$$

$$s_{DSB}(t) = f(t) \cos \omega_C t \overset{\mathcal{F}}{\longleftrightarrow} \frac{1}{2} [F (\omega + \omega_C) + F (\omega - \omega_C)]\qquad(11.4\text{-}2)$$

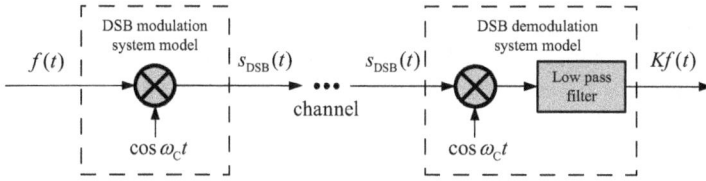

Fig. 11.9: DSB modulation and demodulation system models.

The changes of signals and spectrums in the process of modulation are illustrated in
▶ Figure 11.10.

From ▶ Figure 11.10, it can be seen that the spectrum of the product of the modulating signal and the carrier does not have the impulse component that appeared
in the carrier spectrum. Moreover, on two sides of the carrier frequency w_C, there are
two complete symmetrical frequency spectrums of the modulating signal [this can be
clearly seen from equation (11.4-2)]. The part of the spectrum with frequencies which
are less than the carrier frequency w_C is called the lower sideband spectrum; the part
in the spectrum, whose frequencies are greater than w_C is called the upper sideband
spectrum. This modulating method of which the modulated signal spectrum contains
the upper and the lower sidebands and no impulse components is called the suppressed carrier double sideband amplitude modulation or, simply, DSB modulation.
The modulated signal in DSB usually is written as $s_{DSB}(t)$. At the same time, the amplitude of $s_{DSB}(t)$ changes with the change of the low frequency signal $f(t)$, namely, the

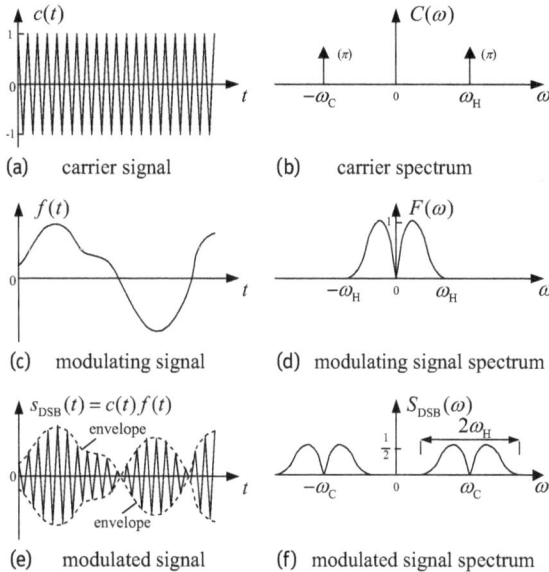

Fig. 11.10: DSB modulation process.

modulating signal seems to be placed on the amplitude of the carrier. From the point of view of the frequency domain, compared with the spectrum of $s_{DSB}(t)$ and the spectrum of $f(t)$, only amplitude values of the former are reduced to half of the latter, and the shape is the same as that of the latter. The former can be considered as equivalent with the latter shifted to place ω_C.

11.4.2 DSB demodulation system

As mentioned before, a modulating system can transform a lower frequency signal into a higher frequency signal. As a result, readers naturally want to know how to recover the lower frequency modulating signal, which is also the original signal from the higher frequency modulated signal.

From ▶ Figure 11.10, we can see that although the amplitude of the modulated signal is controlled by that of the modulating signal, the waveform of the modulated signal is different from the modulating signal's. So, even if the modulated signal has been received by the receiving end, the communication work has not been fulfilled yet; the original (modulating) signal must be extracted from the modulated signal. In other words, it is necessary to demodulate the modulated signal on the receiving end. There is more than one kind of demodulation method, and the coherent method is usually used in the demodulation to DSB signals.

From the trigonometric formula in mathematics,

$$\cos \omega_C t \cdot \cos \omega_C t = \cos^2 \omega_C t = \frac{1}{2} + \frac{1}{2} \cos 2\omega_C t .$$

From the view of communication, the multiplication of two cosine signals in the above formula is a modulation procedure of which a signal (carrier) is modulated by another signal with the same frequency and phase. Then a DC component and a carrier component with twice the carrier frequency can be resulted. This is just the basic principle of coherent demodulation,

$$s_{DSB}(t) \cos \omega_C t = f(t) \cos \omega_C t \cdot \cos \omega_C t = f(t) \cos^2 \omega_C t$$

$$= \frac{1}{2} f(t) + \frac{1}{2} f(t) \cos 2\omega_C t . \tag{11.4-3}$$

This formula states that if the received DSB signal at the receiving end is again modulated by a (local) carrier whose frequency and phase are the same as the sending carrier, the original signal component can appear in the modulated signal. If the term including the carrier component with two times the carrier frequency in (11.4-3) is removed by a lowpass filter, then the remainder is just the original signal term.

This demodulating method, of which the original signal can be obtained by means of the modulated signal being directly multiplied by the local carrier and is then filtered at the receiving end, is called the coherent demodulation or synchronous demodulation. The local carrier is a carrier generated by the receiving end and whose frequency and phase are the same as the sending carrier generated by the sending end. The block diagram of the coherent demodulation is shown in ▸ Figure 11.9.

It should be explained that the coherent demodulation method is more complex because the local carrier is hard to generate by the receiver. If it cannot be guaranteed that the local carrier has the same frequency and phase as that of the sending carrier, the demodulation task will be difficult to complete. The DSB technology is mainly used in stereo radios, color TV systems and other fields.

11.4.3 AM modulation system

In a DSB modulated signal, the amplitude changes with the modulating signal, but its time domain waveform is only the same as a part of modulating signal in the envelope. Specifically, the envelope of the modulated signal is a linear relationship with the waveform of which the modulating signal is rectified in the full wave. Then can we find a way to make the envelope of the modulated signal change linearly with the change of the modulating signal? The answer is "yes". This brings us to the content about the conventional double-band amplitude modulation or, simply, AM modulation.

From ▸ Figure 11.10, if $f(t)$ has no negative values, the envelope of the modulated signal is a linear relationship with the amplitude of the modulating signal $f(t)$. So, how can we change the negative values part of the modulating signal into positive values? The answer is simple; the modulating signal can be added by a constant A whose value is greater than or equal to the minimum negative value of the signal, that is, the modulating signal is shifted up to the A value in the waveform. After the signal is moved up, it is multiplied by the carrier, so the modulated signal whose envelope is a linear relationship with the amplitude changes of the modulating signal can be obtained. Thus, we have the conventional amplitude modulation signal.

The modulating method of which after the original signal is moved up, it is multiplied by a carrier, is just the conventional double sideband amplitude modulation, shorthand for AM.

The modulating procedure is shown in ▸ Figure 11.11. The mathematical deductions are as follows:

If a modulating signal is $f(t)$, its spectrum is $F(\omega)$, and then we have

$$f(t) \overset{\mathcal{F}}{\longleftrightarrow} F(\omega) \,,$$

$$A + f(t) \overset{\mathcal{F}}{\longleftrightarrow} 2\pi A \delta(\omega) + F(\omega) \,.$$

Let the carrier be $c(t)$, and

$$c(t) = \cos \omega_C t \overset{\mathcal{F}}{\longleftrightarrow} \pi \left[\delta \left(\omega + \omega_C \right) + \delta \left(\omega - \omega_C \right) \right] \,.$$

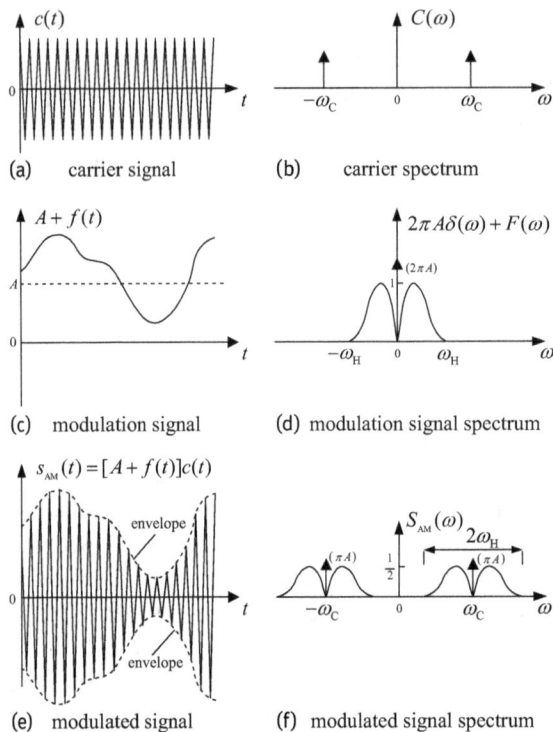

(a) carrier signal

(b) carrier spectrum

(c) modulation signal

(d) modulation signal spectrum

(e) modulated signal

(f) modulated signal spectrum

Fig. 11.11: AM modulation process.

Then modulated signal is

$$s_{AM}(t) = [A + f(t)] \cos \omega_C t = A \cos \omega_C t + f(t) \cos \omega_C t . \qquad (11.4\text{-}4)$$

Its spectrum is

$$S_{AM}(\omega) = \frac{1}{2\pi} \{ [2\pi A \delta(\omega) + F(\omega)] * \pi [\delta(\omega + \omega_C) + \delta(\omega - \omega_C)] \}$$

$$= [2\pi A \delta(\omega) + F(\omega)] * \frac{1}{2} [\delta (\omega + \omega_C) + \delta (\omega - \omega_C)] \qquad (11.4\text{-}5)$$

$$= \pi A [\delta (\omega + \omega_C) + \delta (\omega - \omega_C)] + \frac{1}{2} [F (\omega + \omega_C) + F (\omega - \omega_C)] .$$

Comparing equation (11.4-2) with equation (11.4-4), we can see that the AM signal is more than a DSB signal with a carrier term $A \cos \omega_C t$.

11.4.4 AM demodulation system

Comparing the spectrums and the time domain waveforms of DSB and AM signals, we find that the AM signal is more than the DSB signal with a carrier component, namely,

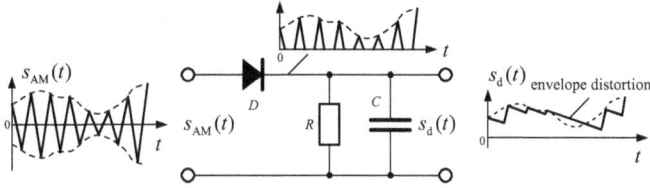

Fig. 11.12: Envelope demodulator.

an impulse signal in the spectrum. This means that the AM signal is more than a DSB signal to send a carrier component; namely, if they have the same launching power for sideband signals, the whole launching power of an AM signal is greater than that of a DSB signal. So, what is the benefit of using more powers to transmit the carrier component? The answer is that its advantage is reflected in the demodulation.

An AM signal is usually demodulated by the envelope demodulating method rather than the coherent way. The envelope demodulator (in ▶ Figure 11.12) is very simple; usually has only the three components: diode D, a capacitor C and a resistor R. The demodulating principle of it is that the envelope of signal $s_{AM}(t)$ can be extracted by the halfwave rectification of diode and the charge–discharge characteristic of the capacitor, but it needs a precondition that the capacitor discharge time is much slower than the charge time.

It needs to be explained that the output waveform of the diode and the voltage waveform of the capacitor should be the same. In ▶ Figure 11.12, the influence of the capacitor is not considered for the diode output waveform; the purpose is to illustrate the demodulation principle.

Of course, an AM signal can also be demodulated by a coherent demodulating method like the DSB signal. ▶ Figure 11.13 gives an AM signal transmission system model of using the coherent demodulator. The AM system is widely used in the radio field usually is employed by the medium-wave and short wave radio broadcasting systems we are familiar with.

It should be noted that the signals and systems knowledge in practical applications is far greater than what has been discussed above, for example, SSB modulation, VSB modulation, frequency division multiplexing, Laplace transform applications in control systems, and usages of discrete signals and systems are not involved in this

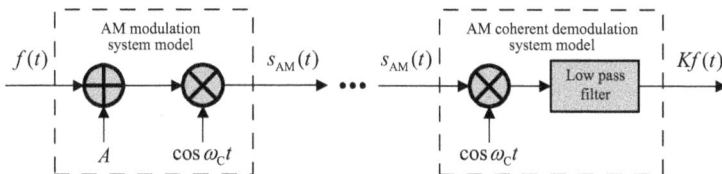

Fig. 11.13: AM modulation and demodulation system model.

chapter. The interested reader is referred to "Communication Principle", "Automatic Control Principle" and "Digital Signal Processing" and other books. In addition, the DSB, SSB and AM modulations are all methods where a modulating signal is modulated onto the amplitude parameter of a carrier, so they all can be called amplitude modulation.

Example 11.4-1. A signal processing system and the frequency characteristic of an ideal lowpass filter $H(j\omega) = G_{2\omega_0}(\omega)$ are separately plotted in ▸ Figure 11.14a and b. Knowing $f(t) = 2 \cos \omega_m t$, $c(t) = \cos \omega_0 t$ and $\omega_m \ll \omega_0$, find $y(t)$.

Solution. According to the Fourier transform,

$$F(j\omega) = 2\pi \left[\delta \left(\omega - \omega_m \right) + \delta \left(\omega + \omega_m \right) \right] ,$$
$$C(j\omega) = \pi \left[\delta \left(\omega - \omega_0 \right) + \delta \left(\omega + \omega_0 \right) \right] .$$

Their waveforms are shown in ▸ Figure 11.14c.
 Because $y_1(t) = f(t)c(t)$,

$$Y_1 (j\omega) = \frac{1}{2\pi} F (j\omega) * C (j\omega)$$
$$= \pi \{\delta \left[\omega - (\omega_0 + \omega_m) \right] + \delta \left[\omega + (\omega_0 + \omega_m) \right]$$
$$+ \delta \left[\omega - (\omega_0 - \omega_m) \right] + \delta \left[\omega + (\omega_0 - \omega_m) \right] \} .$$

Its waveform is shown in ▸ Figure 11.14d.
 From the system diagram, the spectrum of $y(t)$ from the lowpass filter is

$$Y(j\omega) = Y_1 (j\omega) H (j\omega) = \pi \{\delta \left[\omega - (\omega_0 - \omega_m) \right] + \delta \left[\omega + (\omega_0 - \omega_m) \right] \} .$$

Its waveform is shown in ▸ Figure 11.14d. The inverse Fourier transform of $Y(j\omega)$ is

$$y(t) = \mathcal{F}^{-1} [Y(j\omega)] = \cos(\omega_0 - \omega_m)t .$$

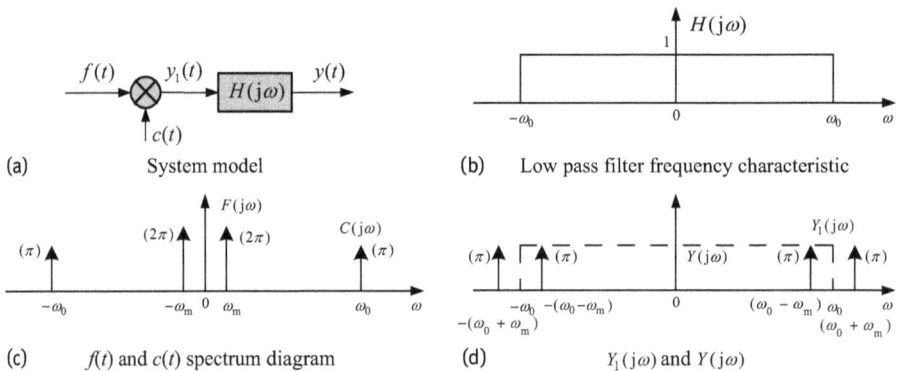

(a) System model

(b) Low pass filter frequency characteristic

(c) $f(t)$ and $c(t)$ spectrum diagram

(d) $Y_1(j\omega)$ and $Y(j\omega)$

Fig. 11.14: E11.4-1.

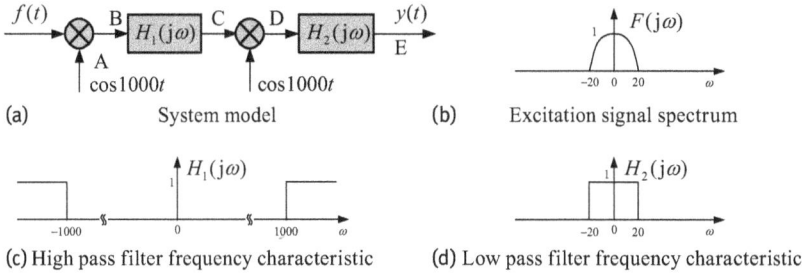

(a) System model

(b) Excitation signal spectrum

(c) High pass filter frequency characteristic

(d) Low pass filter frequency characteristic

Fig. 11.15: E11.4-2.

From this example, we can find that the output of the multiplier is a DSB signal and the lowpass filter outputs a signal only including the lower sideband part in $Y_1(j\omega)$, which is called the SSB signal. The system is equivalent to a frequency converter changing a low frequency signal $s(t) = 2\cos\omega_m t$ into a high frequency signal $y(t) = \cos(\omega_0 - \omega_m)t$.

Example 11.4-2. A signal processing system, $H_1(j\omega)$, $H_2(j\omega)$ and the frequency characteristic of excitation $f(t)$ are as shown in ▶ Figure 11.15a–d.
1. Draw the waveforms of points A, B, C, D in the frequency domain and the spectrum of response $y(t)$.
2. Find the relationship between $y(t)$ and $f(t)$.

Solution. Setting the signal spectrum as $F_A(j\omega)$ at point A, we have

$$F_A(j\omega) = \pi[\delta(\omega + 1\,000) + \delta(\omega - 1\,000)] .$$

The frequency domain waveform at point A is shown in ▶ Figure 11.16a.
Setting the signal spectrum as $F_B(j\omega)$ at point B, from the modulation characteristic, we have

$$F_B(j\omega) = \frac{1}{2\pi}F(j\omega) * F_A(j\omega) = \frac{1}{2}[F_A(j(\omega + 1\,000)) + F_A(j(\omega - 1\,000))] .$$

The frequency domain waveform of point B is shown in ▶ Figure 11.16b.
Setting the signal spectrum as $F_C(j\omega)$ at point C,

$$F_C(j\omega) = F_B(j\omega)H_1(j\omega) .$$

The frequency domain waveform at point C is shown in ▶ Figure 11.16c. Obviously, the high pass filter outputs an SSB signal with only the upper sideband.
Setting the signal spectrum as $F_D(j\omega)$ at point D,

$$F_D(j\omega) = \frac{1}{2\pi}F_C(j\omega) * \pi[\delta(\omega + 1\,000) + \delta(\omega - 1\,000)]$$
$$= \frac{1}{2}[F_C(j(\omega + 1\,000)) + F_C(j(\omega - 1\,000))] .$$

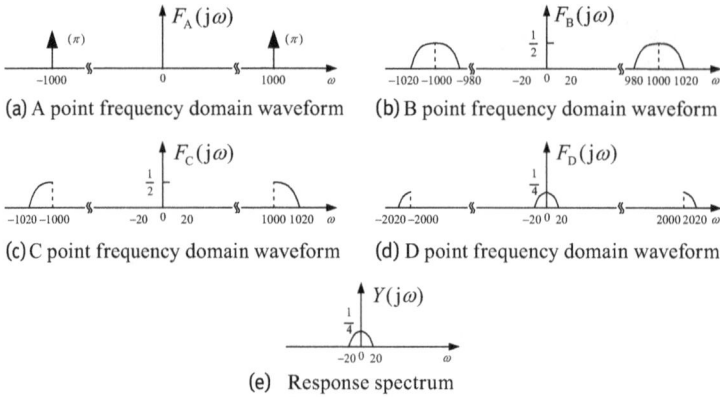

(a) A point frequency domain waveform

(b) B point frequency domain waveform

(c) C point frequency domain waveform

(d) D point frequency domain waveform

(e) Response spectrum

Fig. 11.16: E11.4-2.

The frequency domain waveform at point D is shown in ▸ Figure 11.16d.
Finally, the spectrum of the low pass filter output $y(t)$ is

$$Y(j\omega) = F_D(j\omega)H_2(j\omega) = \frac{1}{4}F(j\omega) .$$

Its waveform is shown in ▸ Figure 11.16e.
Obviously, the relationship between $y(t)$ and $f(t)$ is

$$y(t) = \frac{1}{4}f(t) .$$

The important meaning of this example is that the working principles of common modulating and demodulating systems in communication engineering are illustrated. The first multiplier and the high pass filter are composed of a modulating system at the sending end, which can transfer a low frequency signal into a high frequency signal. That is, the waveform at point C; the signal at point C arrives the input port of the demodulating system at receiving end through the channel (wireless channel). The port is also the input port of second multiplier, and then the signal at the port is modulated by the multiplier again and becomes the sum of a low frequency signal and a higher frequency signal. The synthetic signal passes the lowpass filter to result in the low frequency signal $y(t)$, which is a linear relation with the original signal $f(t)$ at the sending end. That is, the original signal $f(t)$ has been restored at the receiving end, and $y(t)$ and $f(t)$ only have a multiple in amplitudes.

11.5 Solved questions

Question 11-1. In the following systems, the system the signal $e(t) = \cos(10t) \cdot \cos(1\,000t)$ passing through without distortion is ( ).

A. $H(s) = \frac{s+3}{(s+1)(s+2)}$

B. $H(s) = \frac{(s-1)(s-2)}{(s+1)(s+2)}$

C. $H(j\omega) = [\varepsilon(\omega + 1\,100) - \varepsilon(\omega - 1\,100)]e^{-5j\omega}$

D. $h(t) = Sa(5t)\cos(1\,000t)$

Solution. Choose C. Because the signal $e(t)$ is distributed at $[-1\,010, +1\,010]$ in the frequency domain, it is all in the frequency band of the system C, and the system C has a linear phase characteristic, so the signal will not be distorted in system C.

Question 11-2. If a band limited signal $f(t)$ is sampled at the Nyquist rate $(2f_m\,\text{Hz})$. Prove that $f(t)$ can be expressed by the sampled signal $f_s(t) = \sum_{n=-\infty}^{\infty} f(nT)Sa[(\pi/T)(t-nT)]$, where $T = 1/(2f_m)$ is the sampling interval.

Proof. The signal $f_s(t)$ can be expressed as the product of $f(t)$ and the impulse train, namely,

$$f_s(t) = f(t) \sum_{n=-\infty}^{\infty} \delta(t - nT) = \sum_{n=-\infty}^{\infty} f(nT)\delta(t - nT).$$

If the Fourier transforms of $f(t)$ and $\delta(t-nT)$ are $F(\omega)$ and $\delta_T(\omega)$, the Fourier transform of $f_s(t)$ is

$$F_s(\omega) = \frac{1}{2\pi}F(\omega) * \delta_T(\omega) = \frac{1}{2\pi}F(\omega) * \frac{2\pi}{T}\sum_{n=-\infty}^{\infty} \delta\left(\omega - n\frac{2\pi}{T}\right)$$

$$= \frac{1}{T}\sum_{n=-\infty}^{\infty} F\left[\left(\omega - n\frac{2\pi}{T}\right)\right].$$

If $n = 0$, $F_s(\omega) = \frac{1}{T}F(\omega)$, so the spectrum of $f(t)$ can be extracted from a lowpass filter,

$$F(\omega) = TF_s(\omega) \cdot g_{2\pi/T}(\omega).$$

In the formula, $g_{2\pi/T}(\omega)$ is the frequency characteristic of the lowpass filter. Taking the Fourier inverse transform to the formula, we can obtain

$$f(t) = Tf_s(t) * \mathcal{F}^{-1}[g_{2\pi/T}(\omega)].$$

Because the gate signal and the sampling signal are the Fourier transform pair, namely,

$$\mathcal{F}^{-1}[g_{2\pi/T}(\omega)] = \frac{1}{T}Sa\left(\frac{\pi t}{T}\right),$$

and then

$$f(t) = T\sum_{n=-\infty}^{\infty} f(nT)\delta(t - nT) * \left[\frac{1}{T}Sa\left(\frac{\pi t}{T}\right)\right] = \sum_{n=-\infty}^{\infty} f(nT)Sa\left[\frac{\pi}{T}(t - nT)\right].$$

Note: This answer is the proof of the sampling theorem. In addition, the signal spectrum and channel frequency characteristic are commonly expressed by $F(\omega)/F(f)$ and $H(\omega)/H(f)$ in the communication technology field. □

Question 11-3. An ideal low pass filter $h(t) = \frac{\sin 5\pi(t-4)}{\pi(t-4)}$, when $x(t) = \sin \pi t$, output $y(t) = (\quad)$.

A. $y(t) = \sin \pi t$ C. $y(t) = \sin \pi(t - 4)$

B. $y(t) = \sin 5\pi t$ D. $y(t) = \sin 5\pi(t - 4)$

Solution. Choose A. Because the Fourier transform of $h(t) = \frac{\sin 5\pi(t-4)}{\pi(t-4)}$ is $\varepsilon(\omega + 5\pi) - \varepsilon(\omega - 5\pi)$, while the Fourier transform of $x(t) = \sin \pi t$ is $j\pi[\delta(\omega + \pi) - \delta(\omega - \pi)]$, its frequency band is within the range of $|\omega| < \pi$, so the output is $y(t) = \sin \pi t$.

Question 11-4. The system shown in ▸ Figure Q11-4 is a band stop filter composed with two lowpass filters.

(1) If $H_1(j\omega)$ and $H_2(j\omega)$ are the ideal lowpass filter, which has cutoff frequency $\omega_{c1} = 3\pi$ and $\omega_{c2} = \pi$, respectively, namely

$$H_1(j\omega) = \begin{cases} 1 & |\omega| < \omega_{c1} \\ 0 & |\omega| > \omega_{c1} \end{cases}, H_2(j\omega) = \begin{cases} 1 & |\omega| < \omega_{c2} \\ 0 & |\omega| > \omega_{c2} \end{cases},$$

prove that the whole system is equivalent to an ideal band stop filter, and solve the unit impulse response $h(t)$ of the filter.

(2) If input $x(t) = 1 + 2\cos 2\pi t + \sin 4\pi t$, find the output $y(t)$ of the system.

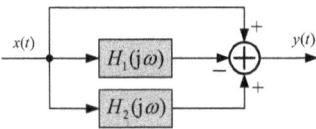

Fig. Q11-4

Solution. (1) The system function of the whole system is

$$H(j\omega) = 1 - H_1(j\omega) + H_2(j\omega) = \begin{cases} 1 & |\omega| < \pi \quad \text{or} \quad |\omega| > 3\pi \\ 0 & \text{other} \end{cases}.$$

So $H(j\omega)$ is the ideal band stop filter characteristic shown in ▸ Figure Q11-4 (1). Obviously, $H(j\omega)$ can be rewritten as $H(j\omega) = 1 - \hat{H}(j\omega)$, where

$$\hat{H}(j\omega) = \begin{cases} 1 & \pi < |\omega| < 3\pi \\ 0 & \text{other} \end{cases}$$

can be considered as the difference of two different gate signals with different widths. Therefore, the inverse transform of $H(j\omega)$ is

$$h(t) = \mathcal{F}^{-1}\{H(j\omega)\} = \delta(t) - \frac{\sin 3\pi t}{\pi t} + \frac{\sin \pi t}{\pi t}.$$

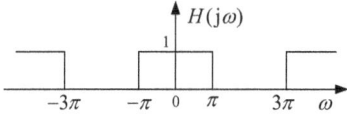

Fig. Q11-4 (1)

(2) Because the system is an ideal band stop filter, the signal components in the frequency band range $\pi < |\omega| < 3\pi$ will be filtered out. However, $x(t) = 1+2\cos 2\pi t + \sin 4\pi t$ composed of three frequency components, the frequency of first term "1" is 0, the frequency of second term "$2\cos 2\pi t$" is 2π, the frequency of third term "$\sin 4\pi t$" is 4π, and then, the term $2\cos 2\pi t$ will be filtered out, so the output is $y(t) = 1 + \sin 4\pi t$.

11.6 Learning tips

The ultimate goal of learning signals and systems analysis is to apply the theory to practice, so it is suggested that readers should focus on the following points:
1. The Fourier series and the Fourier transform are widely used in communication systems because of their physical meaning. The frequency spectrum, power spectrum and energy spectrum of a signal, and the frequency feature of system are all very important.
2. We can use the example that a truck is as a signal and a station is as a system to deepen understanding concepts of a signal passing a system and a system processing a signal.
3. The gate signal and the sample signal are a pair of good partners that play important roles in the communication engineering field, and they almost always come in pairs.

11.7 Problems

Problem 11-1. A system is shown in ▶ Figure P11-1. $f(t) = \sum_{n=-\infty}^{\infty} e^{j2nt}$, $c(t) = \cos 2t$. If

$$H(j\omega) = \begin{cases} \frac{1}{2}, & |\omega| < 3 \text{ rad/s} \\ 0, & |\omega| > 3 \text{ rad/s} \end{cases},$$

find the output $y(t)$.

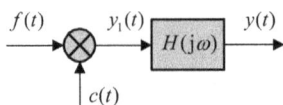

Fig. P11-1

Problem 11-2. The impulse response of a CT system is $h(t) = \frac{1}{\pi}Sa(3t)$, and the input is $f(t) = 3 + \cos 2t$. Find the steady state response $y(t)$ of the system.

Problem 11-3. A communication system is shown in ▶ Figure P11-3, the input is $f(t)$, the highest frequency is ω_H, and $c(t) = A \cos \omega_0 t$, $\omega_H \ll \omega_0$. Expect the output of system to be $y(t) = f(t)$. Find the system function $H(j\omega)$ of the low pass filter (LF) and explain the role of the system.

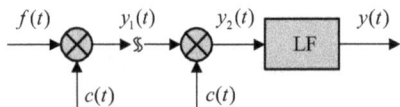

Fig. P11-3

Problem 11-4. A DSB demodulation system in practice is shown in ▶ Figure P11-4a, the amplitude spectrum of the system function of the low pass filter in it is shown in ▶ Figure P11-4b. The phase spectrum $\varphi(\omega) = 0$ and $s(t) = \cos(1\,000t)$ $(-\infty < t < \infty)$. If the input signal is $f(t) = \frac{\sin t}{\pi t} \cos(1\,000t)$ $(-\infty < t < \infty)$, find the system output signal $y(t)$.

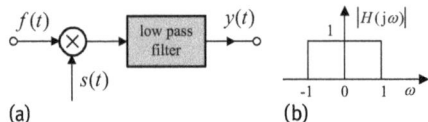

(a)   (b)   **Fig. P11-4**

A Reference answers

Chapter 8

Problem 8-1:

(1) yes, $T = 8$ (2) no (3) yes, $T = 8$ (4) no

Problem 8-2:

(1)–(3) are omitted.

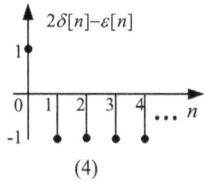

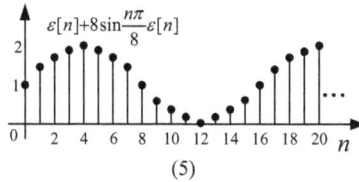

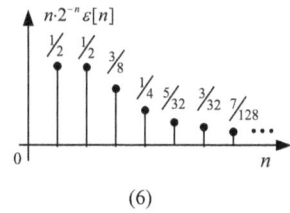

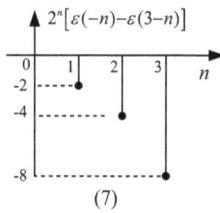

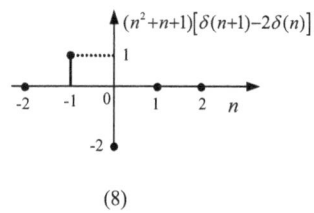

Fig. A8-2

Problem 8-3:

(2) and (3) are omitted.

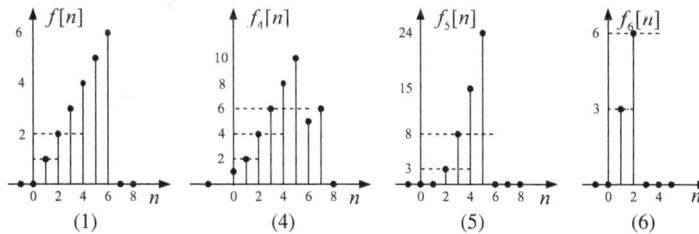

Fig. A8-3

https://doi.org/10.1515/9783110541205-app-005

Problem 8-4:

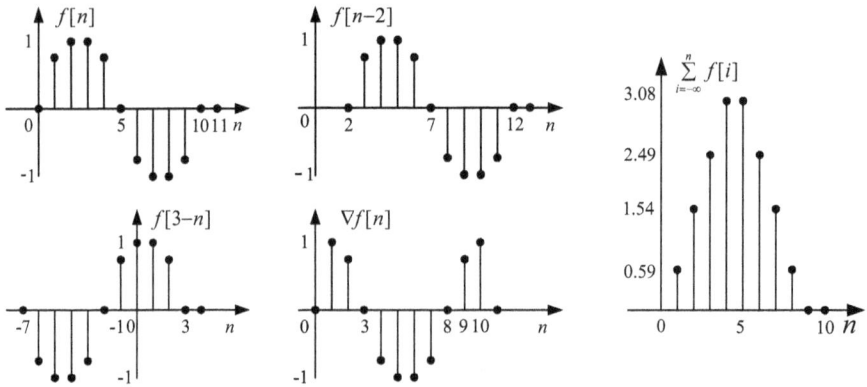

Fig. A8-4

Problem 8-5:

(1) $\delta[n]$

(2) $\varepsilon[n-1]$

(3) $(2n-1)\varepsilon[n-1]$

(4) $\delta[n] + a^{n-1}(a-1)\varepsilon[n-1]$

Problem 8-6:

(a) $f[n] = 2\left[\varepsilon[n] - \varepsilon[n-5]\right]$

(b) $f[n] = \frac{1}{2}(n+1)\varepsilon[n]$

(c) $f[n] = -\left[\varepsilon[n-1] - \varepsilon[n-4]\right] + \left[\varepsilon[n-1] - \varepsilon[-n-4]\right]$

Problem 8-7:

(1) $y[n] = (n+1)\varepsilon[n]$

(2) $y[n] = 2\left[1 - \left(\frac{1}{2}\right)^{n+1}\right]\varepsilon[n]$

(3) $y[n] = 3\delta[n] + 5\delta[n-1] + 6\delta[n-2] + 6\delta[n-3] + 3\delta[n-4] + \delta[n-5]$

(4) $y[n] = (2 - 0.5^n)\varepsilon[n]$

Problem 8-8:

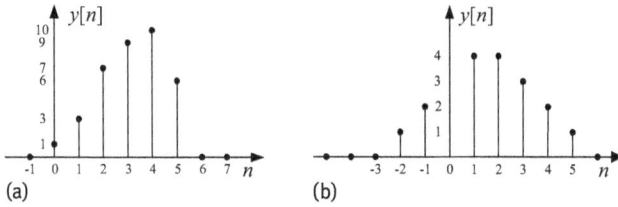

(a)   (b)

Fig. A8-8

Problem 8-9:
The answer is omitted.

Problem 8-10:
$f_1[n] * f_2[n] = \{\ldots, 0, 1, 3, 6, 6, 6, 5, 3, 0, \ldots\}$
$\uparrow$

Problem 8-11:
$y[n] = \left[4 + 3\left(\frac{1}{2}\right)^n - \left(-\frac{1}{2}\right)^n\right]\varepsilon[n]$

Problem 8-12:
(1) $y[n] - 7y[n-1] + 10y[n-2] = 14f[n] - 85f[n-1] + 111f[n-2]$
(2) $y_f[n] = 2(2^n + 3 \times 5^n + 10)\varepsilon[n] - 2\left[2^{(n-10)} + 3 \times 5^{(n-10)} + 10\right]\varepsilon[n-10]$

Problem 8-13:
(1) $H(E) = \frac{E}{E-2}$, $h[n] = 2^n\varepsilon[n]$
(2) $H(E) = \frac{E^2+2E}{E^2-7E+10}$, $h[n] = \left(-\frac{4}{3}\right)2^n\varepsilon[n] + \left(\frac{7}{3}\right)5^n\varepsilon[n]$
(3) $H(E) = \frac{E}{E^2+3E+2}$, $h[n] = (-1)^n\varepsilon[n] - (-2)^n\varepsilon[n]$

Problem 8-14:
(1) $y[n] = -\frac{26}{35}\left(-\frac{1}{3}\right)^n + \frac{6}{5}\left(\frac{1}{2}\right)^n$ $n \geq 0$
(2) $y[n] = (-1)^{n+1} + 2(-2)^n$ $n \geq 0$
(3) $y[n] = (1 + \frac{1}{2}n)\cos\frac{\pi}{2}n$ $n \geq 0$

Problem 8-15:

(1) $h[n] = \delta[n] + 10\left(-\frac{1}{3}\right)^n \varepsilon[n-1]$;

(2) $h[n] = [(-1)^n - (-2)^n]\varepsilon[n]$

(3) $h[n] = \left(\frac{1}{4}\right)^n \varepsilon[n]$;

(4) $h[n] = [0.8(-0.8)^n + 0.2(0.2)^n]\varepsilon[n]$

(5) $h[n] = \left[(0.8)^{n-1} - (-0.2)^{n-1}\right]\varepsilon[n-1]$;

(6) $h[n] = (-1)^{n-1}\varepsilon[n-1]$

Problem 8-16:

$h[n] = 2\delta[n-1] + \delta[n-2] + 0.5\delta[n-3]$

Problem 8-17:

$h[n] = \delta[n-1] + (-1)^{n-1}\varepsilon[n-1]$ $g[n] = \left[\frac{3}{2} + \frac{1}{2}(-1)^n\right]\varepsilon[n-1]$

Problem 8-18:

The answer is omitted.

Problem 8-19:

(1) $y_x[n] = 2^{n+1} - 2 \cdot 4^n$ $n \geq -1$,   (3) $y_x[n] = 24\left(\frac{1}{2}\right)^n - 9\left(\frac{1}{3}\right)^n$ $n \geq 0$

(2) $y_x[n] = \left(\frac{6}{5}\right)^n$ $n \geq -1$

Problem 8-20:

(1) $y[n] = \frac{9}{5}2^n + \frac{49}{20}(-3)^n - \frac{1}{4}$;

(2) $y[n] = (0.5)^{n+1} + 0.5(-0.4)^n + 2, n \geq 0$

(3) $y[n] = \frac{1}{1+2e}\left[(1+e)(-2)^{n+1} + e^{-n}\right]$;

(4) $y[n] = \frac{-e}{1+2e}\left[(-2)^{n+1} - e^{-(n+1)}\right]\varepsilon[n]$

Problem 8-21:

$y_x[n] = [6(0.5)^n + (0.2)^n]\varepsilon[n]$, $y_f[n] = 12.5 - [5(0.5)^n + 0.5(0.2)^n]\varepsilon[n]$,
natural or transient response $[(0.5)^n + 0.5(0.2)^n]\varepsilon[n]$,
forced or steady state response 12.5.

Chapter 9

Problem 9-1:

(1) $F(z) = \frac{z}{z+1};\quad p = -1, z = 0, |z| > 1$

(2) $F(z) = \frac{z}{z^2-1};\quad |z| > 1, p = \pm 1, z = 0$

(3) $F(z) = z^{-N},\quad$ except point of origin whole z plane

(4) $F(z) = z^N,\quad$ except point ∞ whole z plane

(5) $F(z) = \frac{2}{2z-1},\quad |z| > \frac{1}{2}, p = \frac{1}{2}$

(6) $F(z) = \frac{1}{1-2z},\quad |z| > \frac{1}{2}, p = \frac{1}{2}$

(7) $F(z) = \frac{z}{z-\frac{1}{2}},\quad |z| > \frac{1}{2}, z = 0, p_1 = \frac{1}{4}, p_2 = \frac{2}{3}$

(8) $F(z) = \frac{-5z}{12(z-\frac{1}{4})(z-\frac{2}{3})},\quad |z| > \frac{2}{3}, z = 0, p_1 = \frac{1}{4}, p_2 = \frac{2}{3}$

Problem 9-2:

(1) $F(z) = \frac{1}{1-2z},\quad |z| < \frac{1}{2};$

(2) $F(z) = \frac{z}{z-\frac{1}{2}} - \frac{2}{z-2},\quad \frac{1}{2} < |z| < 2$

Problem 9-3:

(1) $F(z) = (1 - z^{-8}) \frac{z}{z-1},\quad |z| > 1;$

(2) $F(z) = \frac{z^2}{z^2+1},\quad |z| > 1$

(3) $F(z) = \frac{4z^2}{4z^2+1},\quad |z| > \frac{1}{2};$

(4) $F(z) = \frac{z}{(z-1)^2},\quad |z| > 1$

(5) $F(z) = \frac{2z}{(z-1)^3},\quad |z| > 1;$

(6) $F(z) = \frac{az}{(z-a)^2},\quad |z| > a$

(7) $F(z) = \frac{2z-z^2}{z^2(z-1)^2},\quad |z| > 1;$

(8) $F(z) = \frac{1}{z(z-1)^2},\quad |z| > 1$

(9) $F(z) = \frac{2-z}{(z-1)^2},\quad |z| > 1;$

(10) $F(z) = \frac{1}{z^7}(z + 1)^2(z^2 + 1)^2$

Problem 9-4:

$f_1[n] = \frac{(-a)^n}{n!},\quad n \geq 0; f_2[n] = \frac{1}{(-n)!}\varepsilon[-n]$

Problem 9-5:

(1) $\{2, 0.5, 1.25, 0.875, \ldots\};$

(2) $\{0, 1, 0, 1, \ldots\};$

(3) $\{1, 3.5, 4.75, 6.375, \ldots\}$

Problem 9-6:

(1) $f[n] = [5 + 5 \cdot (-1)^n] \varepsilon[n];$

(2) $f[n] = \frac{6}{5} \cdot 5^n \varepsilon[n] + (-1)^n \varepsilon[n] + \left(-\frac{1}{5}\right) \delta[n]$

(3) $f[n] = (2n - 1)\varepsilon[n - 1];$

(4) $f[n] = 6\delta[n] - \frac{9}{5}\left(-\frac{1}{2}\right)^n \varepsilon[n] - \frac{16}{5}\left(\frac{1}{3}\right)^n \varepsilon[n]$

Problem 9-7:

(1) $f[n] = 7\delta[n-1] + 3\delta[n-2] - 8\delta[n-10]$

(2) $f[n] = 2\delta[n+1] + 3\delta[n] + 4\delta[n-1]$

(3) $f[n] = (-2)^n\varepsilon[n] - 5(-2)^{n-1}\varepsilon[n-1]$

(4) $f[n] = n \cdot 6^{n-1}\varepsilon[n]$,

(5) $f[n] = \frac{1}{n} \cdot \left(\frac{1}{2}\right)^n \varepsilon[-n-1]$

(6) $f[n] = \delta[n] - \cos\frac{n\pi}{2}\varepsilon[n]$,

(7) $f[n] = 2\sin\frac{n\pi}{6}\varepsilon[n]$

(8) $f[n] = \sqrt{2}\cos\left(\frac{3n\pi}{4} + \frac{\pi}{4}\right)\varepsilon[n]$

Problem 9-8:

(1) $f_1[n] = \frac{2}{3}\delta[n] + \left[\frac{1}{3} \cdot 3^n - \left(\frac{1}{2}\right)^n\right]\varepsilon[n]$

(2) $f_2[n] = \frac{2}{3}\delta[n] - \left[\frac{1}{3} \cdot 3^n - \left(\frac{1}{2}\right)^n\right]\varepsilon[-n-1]$

(3) $f_3[n] = \frac{2}{3}\delta[n] - \left(\frac{1}{2}\right)^n\varepsilon[n] - \frac{1}{3} \cdot 3^n\varepsilon[-n-1]$

Problem 9-9:

(1) $a^{n-2}\varepsilon[n-2]$;

(2) $\frac{1-a^{n+2}}{1-a}\varepsilon[n+1]$;

(3) $\frac{b^{n+1}-a^{n+1}}{b-a}\varepsilon[n]$

(4) $(n-3)\varepsilon[n-3]$;

(5) $\frac{1}{2}(n+1)n\varepsilon[n]$;

(6) $\frac{b}{b-a}\left[a^n\varepsilon[n] + b^n\varepsilon[-n-1]\right]$

Problem 9-10:

(1) $y[n] = \left[\frac{1}{6} + \frac{1}{2}(-1)^n - \frac{2}{3}2^n\right]\varepsilon[n]$

(2) $y[n] = \left[\left(\frac{1}{3}n + \frac{7}{12}\right)(-1)^n + \frac{3}{4} \cdot 3^n\right]\varepsilon[n]$

(3) $y[n] = \left[\frac{1}{3} + \frac{4\sqrt{3}}{3}\sin\frac{2\pi}{3}n + \frac{2}{3}\cos\frac{2\pi}{3}n\right]\varepsilon[n]$

Problem 9-11:

$y_{f1}[n] = \frac{1}{1-a}\varepsilon[n] + \frac{1}{a-1}a^{n+1}\varepsilon[n]$;

$y_{f2}[n] = \frac{e^{j\omega(1+n)}}{e^{j\omega}-a}\varepsilon[n] + \frac{1}{a-e^{j\omega}}a^{n+1}\varepsilon[n]$

Problem 9-12:

(1) $H(z) = \frac{z^2+4z}{z^2+3z+2}$, unstable;

(2) $H(z) = \frac{5z^2}{z^2-4z-5}$, unstable

(3) $H(z) = \frac{z^3+z^2+2z+2}{z^3}$, unstable;

(4) $H(z) = \frac{z+2}{z^2+2z+2}$, unstable

Problem 9-13:

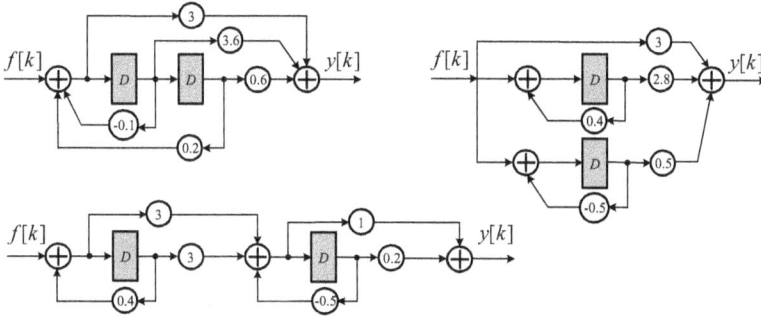

Fig. A9-13 (1)

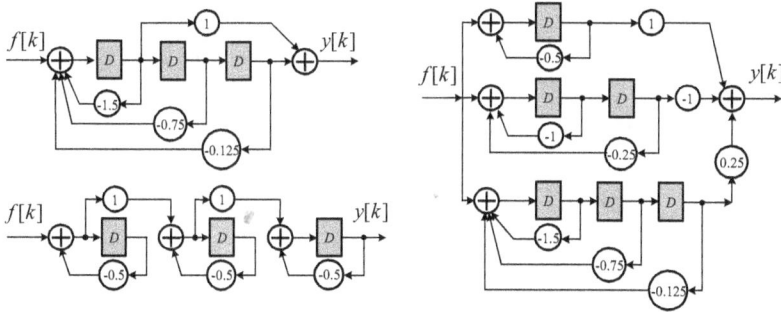

Fig. A9-13 (2)

Problem 9-14:

(1) $H(z) = \frac{1}{z^2+z+k+0.25}$,

(2) $\frac{1}{4} \le k \le \frac{3}{4}$,

(3) $h[n] = \frac{2}{\sqrt{3}} \sin \frac{2\pi}{3}(n-1)\varepsilon[n-1]$

Problem 9-15:

(1) $H(z) = \frac{z(z+1)}{z^2+0.8z-0.2}$,

(2) it is critically stable

(3) $y[n] + 0.8y[n-1] - 0.2y[n-2] = f[n] + f[n-1]$,

(4) $h[n] = (0.2)^n\varepsilon[n]$

(5) $y_x[n] = \left[\frac{5}{3}(0.2)^n - \frac{2}{3}(-1)^n\right]\varepsilon[n]$, $y_f[n] = \left[\frac{5}{4} - \frac{1}{4}(0.2)^n\right]\varepsilon[n]$

Chapter 10

Problem 10-1:

(a)

$$\begin{bmatrix} i_L' \\ u_C' \end{bmatrix} = \begin{bmatrix} -\frac{R_2}{L} & \frac{1}{L} \\ -\frac{R_1+R_2}{R_1 C} & -\frac{1}{R_1 C_3} \end{bmatrix} \begin{bmatrix} i_L \\ u_C \end{bmatrix} + \begin{bmatrix} 0 \\ \frac{1}{R_1 C} \end{bmatrix} f(t) ;$$

(b)

$$\begin{bmatrix} u_C' \\ i_L' \end{bmatrix} = \begin{bmatrix} -\frac{1}{R_1 C} & -\frac{1}{C} \\ \frac{1}{L} & -\frac{R_2}{L} \end{bmatrix} \begin{bmatrix} u_C \\ i_L \end{bmatrix} + \begin{bmatrix} \frac{1}{C} \\ 0 \end{bmatrix} f(t)$$

(c)

$$\begin{bmatrix} i_{L1}' \\ i_{L2}' \\ u_C' \end{bmatrix} = \begin{bmatrix} -\frac{R_1}{L_1} & 0 & -\frac{1}{L_1} \\ 0 & -\frac{R_2}{L_2} & -\frac{1}{L_2} \\ \frac{1}{C} & \frac{1}{C} & 0 \end{bmatrix} \begin{bmatrix} i_{L1} \\ i_{L2} \\ u_C \end{bmatrix} + \begin{bmatrix} \frac{1}{L_1} & 0 \\ 0 & \frac{1}{L_2} \\ 0 & 0 \end{bmatrix} \begin{bmatrix} f_1(t) \\ f_2(t) \end{bmatrix}$$

$$\begin{bmatrix} i_{L1}' \\ i_{L2}' \\ u_C' \end{bmatrix} = \begin{bmatrix} -2 & 0 & -1 \\ 0 & -2 & -1 \\ 0.5 & 0.5 & 0 \end{bmatrix} \begin{bmatrix} i_{L1} \\ i_{L2} \\ u_C \end{bmatrix} + \begin{bmatrix} 1 & 0 \\ 0 & 1 \\ 0 & 0 \end{bmatrix} \begin{bmatrix} f_1(t) \\ f_2(t) \end{bmatrix}$$

Problem 10-2:

$$\begin{bmatrix} \dot{\lambda}_1 \\ \dot{\lambda}_2 \\ \dot{\lambda}_3 \end{bmatrix} = \begin{bmatrix} -2 & 3 & 0 \\ 0 & 0 & 1 \\ -4 & 5 & -3 \end{bmatrix} \begin{bmatrix} \lambda_1 \\ \lambda_2 \\ \lambda_3 \end{bmatrix} + \begin{bmatrix} 1 & 0 \\ 0 & 0 \\ 2 & 3 \end{bmatrix} \begin{bmatrix} f_1 \\ f_2 \end{bmatrix} ,$$

$$\begin{bmatrix} y_1 \\ y_2 \end{bmatrix} = \begin{bmatrix} 1 & 0 & 0 \\ 0 & 1 & 0 \end{bmatrix} \begin{bmatrix} \lambda_1 \\ \lambda_2 \\ \lambda_3 \end{bmatrix} + \begin{bmatrix} 0 & 0 \\ 0 & 0 \end{bmatrix} \begin{bmatrix} f_1 \\ f_2 \end{bmatrix}$$

Problem 10-3:

(a)

$$\begin{bmatrix} \dot{i}_L \\ \dot{u}_C \end{bmatrix} = \begin{bmatrix} \frac{-R_1}{L} & \frac{-1}{L} \\ \frac{1}{C} & \frac{-1}{R_2 C} \end{bmatrix} \begin{bmatrix} i_L \\ u_C \end{bmatrix} + \begin{bmatrix} \frac{R_1}{L} & 0 \\ 0 & \frac{1}{R_2 C} \end{bmatrix} \begin{bmatrix} i_S \\ u_S \end{bmatrix} ,$$

$$\begin{bmatrix} y_1 \\ y_2 \end{bmatrix} = \begin{bmatrix} -R_1 & 0 \\ 0 & 1 \end{bmatrix} \begin{bmatrix} i_L \\ u_C \end{bmatrix} + \begin{bmatrix} R_1 & 0 \\ 0 & -1 \end{bmatrix} \begin{bmatrix} i_S \\ u_S \end{bmatrix}$$

(b)

$$\begin{bmatrix} \dot{i}_L \\ \dot{u}_C \end{bmatrix} = \begin{bmatrix} -1 & 0.5 \\ -2 & -1 \end{bmatrix} \begin{bmatrix} i_L \\ u_C \end{bmatrix} + \begin{bmatrix} 0 \\ 2 \end{bmatrix} f(t), \quad y(t) = \begin{bmatrix} -2 & 1 \end{bmatrix} \begin{bmatrix} i_L \\ u_C \end{bmatrix}$$

Problem 10-4:

(a)

$$\begin{bmatrix} \dot{x}_1 \\ \dot{x}_2 \\ \dot{x}_3 \end{bmatrix} = \begin{bmatrix} -2 & 0 & 0 \\ 5 & -5 & 0 \\ 5 & -4 & 0 \end{bmatrix} \begin{bmatrix} x_1 \\ x_2 \\ x_3 \end{bmatrix} + \begin{bmatrix} 1 \\ 0 \\ 0 \end{bmatrix} f(t) ; \quad y = \begin{bmatrix} 0 & 0 & 1 \end{bmatrix} \begin{bmatrix} x_1 \\ x_2 \\ x_3 \end{bmatrix}$$

(b)

$$\begin{bmatrix} \dot{x}_1 \\ \dot{x}_2 \\ \dot{x}_3 \end{bmatrix} = \begin{bmatrix} 0 & 0 & 0 \\ 0 & -2 & 0 \\ 0 & 0 & -5 \end{bmatrix} \begin{bmatrix} x_1 \\ x_2 \\ x_3 \end{bmatrix} + \begin{bmatrix} 1 \\ 1 \\ 1 \end{bmatrix} f(t) ; \quad y = \begin{bmatrix} \frac{1}{2} & \frac{5}{6} & \frac{4}{3} \end{bmatrix} \begin{bmatrix} x_1 \\ x_2 \\ x_3 \end{bmatrix}$$

Problem 10-5:

$$\begin{bmatrix} \dot{x}_1(t) \\ \dot{x}_2(t) \end{bmatrix} = \begin{bmatrix} -4 & -3 \\ 0 & -3 \end{bmatrix} \begin{bmatrix} x_1(t) \\ x_2(t) \end{bmatrix} + \begin{bmatrix} 0 \\ 1 \end{bmatrix} f(t); \quad y(t) = \begin{bmatrix} 1 & 0 & 0 \end{bmatrix} \begin{bmatrix} x_1(t) \\ x_2(t) \\ x_3(t) \end{bmatrix} + \begin{bmatrix} 0 \end{bmatrix} f(t)$$

Problem 10-6:

(a)

$$\begin{bmatrix} \dot{\lambda}_1(t) \\ \dot{\lambda}_2(t) \end{bmatrix} = \begin{bmatrix} -3 & 0 \\ 2 & -1 \end{bmatrix} \begin{bmatrix} \lambda_1(t) \\ \lambda_2(t) \end{bmatrix} + \begin{bmatrix} 8 & 7 \\ 4 & 1 \end{bmatrix} \begin{bmatrix} f_1(t) \\ f_2(t) \end{bmatrix} ; \quad y(t) = \begin{bmatrix} 2 & 3 \end{bmatrix} \begin{bmatrix} \lambda_1(t) \\ \lambda_2(t) \end{bmatrix}$$

(b)

$$\begin{bmatrix} \dot{\lambda}_1(t) \\ \dot{\lambda}_2(t) \\ \dot{\lambda}_3(t) \end{bmatrix} = \begin{bmatrix} -4 & -0.5 & 4 \\ 0 & -3 & 4 \\ 0 & 0 & -1 \end{bmatrix} \begin{bmatrix} \lambda_1(t) \\ \lambda_2(t) \\ \lambda_3(t) \end{bmatrix} + \begin{bmatrix} 0 \\ 0 \\ 1 \end{bmatrix} f(t); y(t) = \begin{bmatrix} 1 & 0 & 0 \end{bmatrix} \begin{bmatrix} \lambda_1(t) \\ \lambda_2(t) \\ \lambda_3(t) \end{bmatrix}$$

(c)

$$\begin{bmatrix} \dot{\lambda}_1(t) \\ \dot{\lambda}_2(t) \\ \dot{\lambda}_3(t) \end{bmatrix} = \begin{bmatrix} 0 & 1 & 0 \\ 0 & 0 & 1 \\ 0 & -10 & -7 \end{bmatrix} \begin{bmatrix} \lambda_1(t) \\ \lambda_2(t) \\ \lambda_3(t) \end{bmatrix} + \begin{bmatrix} 0 \\ 0 \\ 1 \end{bmatrix} f(t) ; \quad y(t) = \begin{bmatrix} 5 & 5 \end{bmatrix} \begin{bmatrix} \lambda_1(t) \\ \lambda_2(t) \end{bmatrix}$$

Problem 10-7:

(1)

$$\begin{bmatrix} \dot{x}_1 \\ \dot{x}_2 \end{bmatrix} = \begin{bmatrix} 0 & 1 \\ -3 & -4 \end{bmatrix} \begin{bmatrix} x_1 \\ x_2 \end{bmatrix} + \begin{bmatrix} 0 \\ 1 \end{bmatrix} f , \quad y = \begin{bmatrix} 1 & 1 \end{bmatrix} \begin{bmatrix} x_1 \\ x_2 \end{bmatrix}$$

(2)

$$\begin{bmatrix} \dot{x}_1 \\ \dot{x}_2 \\ \dot{x}_3 \end{bmatrix} = \begin{bmatrix} 0 & 1 & 0 \\ 0 & 0 & 1 \\ -2 & -1 & -5 \end{bmatrix} \begin{bmatrix} x_1 \\ x_2 \\ x_3 \end{bmatrix} + \begin{bmatrix} 0 \\ 0 \\ 1 \end{bmatrix} f , \quad y = \begin{bmatrix} 2 & 1 & 0 \end{bmatrix} \begin{bmatrix} x_1 \\ x_2 \\ x_3 \end{bmatrix}$$

Problem 10-8:

(1)

$$\begin{bmatrix} \dot{x}_1(t) \\ \dot{x}_2(t) \end{bmatrix} = \begin{bmatrix} 1 & 0 \\ -12 & -7 \end{bmatrix} \begin{bmatrix} x_1(t) \\ x_2(t) \end{bmatrix} + \begin{bmatrix} 0 \\ 1 \end{bmatrix} f(t), \quad y(t) = \begin{bmatrix} 10 & 3 \end{bmatrix} \begin{bmatrix} x_1(t) \\ x_2(t) \end{bmatrix}, \quad \text{direct form}.$$

(2)

$$\begin{bmatrix} \dot{x}_1(t) \\ \dot{x}_2(t) \end{bmatrix} = \begin{bmatrix} -3 & 1 \\ 0 & -4 \end{bmatrix} \begin{bmatrix} x_1(t) \\ x_2(t) \end{bmatrix} + \begin{bmatrix} 0 \\ 1 \end{bmatrix} f(t), \quad y(t) = \begin{bmatrix} 10 & 3 \end{bmatrix} \begin{bmatrix} x_1(t) \\ x_2(t) \end{bmatrix}, \quad \text{series form}.$$

(3)

$$\begin{bmatrix} \dot{x}_1(t) \\ \dot{x}_2(t) \end{bmatrix} = \begin{bmatrix} -3 & 0 \\ 0 & -4 \end{bmatrix} \begin{bmatrix} x_1(t) \\ x_2(t) \end{bmatrix} + \begin{bmatrix} 1 \\ 1 \end{bmatrix} f(t), \quad y(t) = \begin{bmatrix} 1 & 2 \end{bmatrix} \begin{bmatrix} x_1(t) \\ x_2(t) \end{bmatrix}, \quad \text{parallel form}.$$

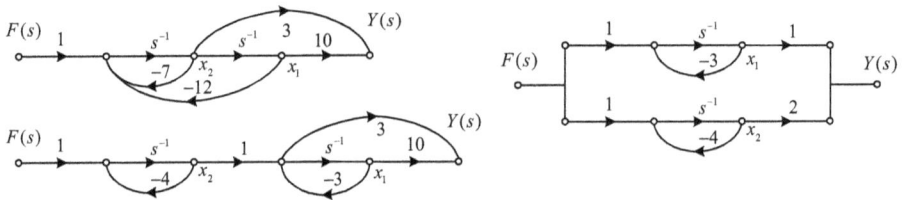

Fig. A10-8

Problem 10-9:

(1)

$$\Phi(t) = \begin{bmatrix} e^{-t}(\cos t + \sin t) & 2e^{-t}\sin t \\ -e^{-t}\sin t & e^{-t}(\cos t - \sin t) \end{bmatrix},$$

$$\alpha_{1,2} = -1 \pm j; \quad \text{system is stable}.$$

(2)

$$\Phi(t) = \begin{bmatrix} 1 & \frac{1}{2}(e^t - e^{-t}) & \frac{1}{2}(e^t + e^{-t} - 1) \\ 0 & \frac{1}{2}(e^t + e^{-t}) & \frac{1}{2}(e^t - e^{-t}) \\ 0 & \frac{1}{2}(e^t - e^{-t}) & \frac{1}{2}(e^t + e^{-t}) \end{bmatrix},$$

$$\alpha_1 = 0, \quad \alpha_2 = -1, \quad \alpha_3 = 1; \quad \text{system is unstable}.$$

(3)

$$\Phi(t) = \begin{bmatrix} e^{-t} & 0 & 0 \\ 0 & e^{2t} & 0 \\ 0 & 0 & e^{-3t} \end{bmatrix},$$

$$\alpha_1 = -1, \quad \alpha_2 = 2, \quad \alpha_3 = -3; \quad \text{system is unstable}.$$

Problem 10-10:

$$y(t) = \left(\frac{1}{2} + e^{-t} - \frac{1}{2}e^{-2t} \right) \varepsilon(t), \quad h(t) = e^{-2t}\varepsilon(t), \quad H(s) = \frac{1}{s+2}$$

Problem 10-11:

(1)

$$A = \begin{bmatrix} -1 & 0 & 0 \\ 0 & -4 & 4 \\ 0 & -1 & 0 \end{bmatrix} ;$$

(2)

$$A = \begin{bmatrix} 0 & 1 \\ -1 & -2 \end{bmatrix}$$

Problem 10-12:

(1)

$$H(s) = \frac{5s+5}{s^3 + 7s^2 + 10s} ;$$

(2)

$$h(t) = \left(\frac{5}{6}e^{-2t} - \frac{4}{3}e^{-5t} + \frac{1}{2} \right) \varepsilon(t)$$

Problem 10-13:

$$y_x(t) = \left(\frac{5}{2}e^{-t} - \frac{15}{2}e^{-3t} \right) \varepsilon(t), \quad y_f(t) = \left(\frac{1}{3} - e^{-t} + \frac{5}{3}e^{-3t} \right) \varepsilon(t)$$

Problem 10-14:

(1)

$$\begin{bmatrix} \dot{x}_1(t) \\ \dot{x}_2(t) \end{bmatrix} = \begin{bmatrix} -4 & 1 \\ -3 & 0 \end{bmatrix} \begin{bmatrix} x_1(t) \\ x_2(t) \end{bmatrix} + \begin{bmatrix} 1 \\ 1 \end{bmatrix} f(t); \quad y(t) = \begin{bmatrix} 1 & 0 \end{bmatrix} \begin{bmatrix} x_1(t) \\ x_2(t) \end{bmatrix}$$

(2)

$$\frac{d^2}{dt^2}y(t) + 4\frac{d}{dt}y(t) + 3y(t) = \frac{d}{dt}f(t) + f(t) ;$$

(3)

$$\begin{cases} x_1(0_-) = 0 \\ x_2(0_-) = 1 \end{cases}$$

Problem 10-15:

(1)

$$\Phi(t) = \begin{bmatrix} 2e^{-t} - e^{-2t} & 2e^{-t} - 2e^{-2t} \\ -e^{-t} + e^{-2t} & -e^{-t} + 2e^{-2t} \end{bmatrix} ;$$

(2)

$$A = \begin{bmatrix} 0 & 2 \\ -1 & -3 \end{bmatrix}$$

Problem 10-16:

(1)

$$\begin{bmatrix} \dot{x}_1(t) \\ \dot{x}_2(t) \\ \dot{x}_3(t) \end{bmatrix} = \begin{bmatrix} 0 & 1 & 0 \\ -2 & -3 & 0 \\ 3 & 1 & -3 \end{bmatrix} \begin{bmatrix} x1(t) \\ x2(t) \\ x3(t) \end{bmatrix} + \begin{bmatrix} 0 \\ 1 \\ 0 \end{bmatrix} f_1(t); \quad y(t) = \begin{bmatrix} 0 & 0 & 1 \end{bmatrix} \begin{bmatrix} x_1(t) \\ x_2(t) \\ x_3(t) \end{bmatrix}$$

$$\begin{bmatrix} \dot{x}_1(t) \\ \dot{x}_2(t) \\ \dot{x}_3(t) \end{bmatrix} = \begin{bmatrix} -3 & 0 & 0 \\ 0 & 0 & 1 \\ 1 & -2 & -3 \end{bmatrix} \begin{bmatrix} x_1(t) \\ x_2(t) \\ x_3(t) \end{bmatrix} + \begin{bmatrix} 1 \\ 0 \\ 0 \end{bmatrix} f_2(t); \quad y(t) = \begin{bmatrix} 0 & 3 & 1 \end{bmatrix} \begin{bmatrix} x_1(t) \\ x_2(t) \\ x_3(t) \end{bmatrix}$$

(2)

$$H_1(s) = \frac{s+3}{(s+1)(s+2)}, \quad H_2(s) = \frac{1}{s+3},$$

$$H_{12}(s) = H_{21}(s) = \frac{1}{(s+1)(s+2)}$$

Problem 10-17:

$$H(s) = \frac{b_2 s^2 + b_1 s + b_0}{s^2 + a_1 s + a_0}$$

Problem 10-18:

$$|sI - A| = \begin{bmatrix} s+3 & 2 \\ -1 & s \end{bmatrix} = s^2 + 3s + 2, \text{ system is stable .}$$

Chapter 11

Problem 11-1:

$y(t) = 1 + 2\cos 2t$

Problem 11-2:

$y(t) = 1 + \frac{1}{3}\cos 2t$

Problem 11-3:

$H(j\omega) = \frac{2}{A^2} G_{2\omega_C}(\omega)$, ω_C is cut off frequency of a gate, $\omega_H \leq \omega_C$. A DSB modulation and demodulation system.

Problem 11-4:

$y(t) = \frac{1}{2\pi} Sa(t)$

Bibliography

Weigang Zhang, Wei-Feng Zhang. Signals and Systems (Chinese edition). Beijing, Tsinghua university press, 2012.

Weigang Zhang. Circuits Analysis (Chinese edition). Tsinghua university press, January 2015.

Alin V. Oppenheim, Alin S. Willsky, S. Hamid Nawab. Signals and Systems (Second Edition). Beijing, Electronic industry press, February 2004.

Bernd Girod, Rudolf Rabenstein, Alexander Strenger. Signals and Systems. Beijing, Tsinghua university press, March 2003.

Charles L. Phillips, John M. Parr, Eve A. Riskin. Signals, Systems and Transforms (Third Edition). Beijing, Mechanical industry press, January 2004.

M. J. Roberts. Signals and Systems – Analysis Using Transform Methods and MATLAB. McGrawHill Companies, 2004.

Weigang Zhang, Lina Cao. Communication Principles (Chinese Edition). Beijing, Tsinghua university press, 2016.

https://doi.org/10.1515/9783110541205-006

Resume

Weigang Zhang, male, Chinese, Professor, Ph.D., graduated from Xidian University in 1982 and used to be an army officer and an engineer in several electrical companies. Teaching for more than 15 years: "Circuits Analysis", "Signals and Systems", "Communication Principles" and other courses such as "Computer Networks", "Introduction to Computer Science" and "ITS Overview". Has written five books in Chinese, e.g., "Circuits Analysis Course", "Signals and Systems Course" and "Communication Principle Course". Currently employed by Chang'an University as the Head of the Department of IoT and Network Engineering and Director of the Highway Traffic Computer Technology Institute.

https://doi.org/10.1515/9783110541205-007

Index

https://doi.org/10.1515/9783110541205-008